Citrus Research - Horticultural and Human Health Aspects

Edited by Mateus Pereira Gonzatto and Júlia Scherer Santos

Published in London, United Kingdom

Citrus Research - Horticultural and Human Health Aspects
http://dx.doi.org/10.5772/intechopen.100891
Edited by Mateus Pereira Gonzatto and Júlia Scherer Santos

Contributors
Zeinab Rafie-Rad, Majid Moradkhani, Ahmad Golchin, Taqi Raza, Neal S. Eash, Anne Adebukola Adeyanju, Muhammad Babar Shahzad Afzal, Ansa Banazeer, Jose Eduardo Serrao, Muhammad Rizwan, Afifa Naeem, Abduljelili Uthman, Yahaya Garba, Juan Carlos Caicedo, Sonia Villamizar, Faezeh Falaki, Angel Villegas - Monter, Itzel Villegas-Velazquez, Maria Andrade-Rodriguez, Elisa Del Carmen Martinez-Ochoa, Alejandra Salvador, Julia Morales, Lourdes Cervera, Pilar Navarro, Sakhawat Riaz, Arslan Ahmad, Rimsha Farooq, Nasir Hussain, Tariq Riaz, Khadim Hussain, Muhammad Mazahir, Mateus Pereira Pereira Gonzatto, Júlia Scherer Santos

First published in London, United Kingdom, 2023 by IntechOpen
IntechOpen is the global imprint of INTECHOPEN LIMITED, registered in England and Wales, registration number: 11086078, 5 Princes Gate Court, London, SW7 2QJ, United Kingdom

British Library Cataloguing-in-Publication Data
A catalogue record for this book is available from the British Library

Additional hard and PDF copies can be obtained from orders@intechopen.com

Citrus Research - Horticultural and Human Health Aspects
Edited by Mateus Pereira Gonzatto and Júlia Scherer Santos
p. cm.
Print ISBN 978-1-80355-804-2
Online ISBN 978-1-80355-805-9
eBook (PDF) ISBN 978-1-80355-806-6

Meet the editors

Dr. Mateus Pereira Gonzatto is a Professor of Citrus and Mango Ecophysiology, in the Department of Agronomy at the Federal University of Viçosa, Brazil. He is a director of representations and publications of the Brazilian Fruit Growing Society (SBF). He researches topics such as citriculture, scion/interstock/rootstock interactions, plant growth regulators, biostimulants, fruit thinning, fruit maturation, and orchard management.

Júlia Scherer Santos is a pharmacist with a Ph.D. in Pharmaceutical Nanotechnology from the Federal University of Rio Grande do Sul, Brazil. She is a specialist in cosmetology and aesthetic health. She works in the areas of cosmetology, nanotechnology, pharmaceutical technology, beauty, and aesthetics. Dr. Santos also has experience as a professor of graduate courses.

Contents

Preface

This book gathers the state of the art of several aspects of citrus research. It consists of eleven chapters grouped into three sections: "Citrus Physiology and Production Technology", "Citrus Pests and Disease", and "Citrus in Human Health".

The first section discusses aspects of physiology and management of citrus production. Chapter 1 provides a review of world citrus production over the last five years. Chapter 2 examines different abiotic stresses in citrus crops and strategies for mitigating them in citrus production. Chapter 3 discusses citrus polyembryony, a relevant theme regarding citrus rootstocks. Chapter 4 discusses the degreening of citrus fruits and its effect on postharvest quality. Chapter 5 reviews citrus mineral nutrition and its consequences for both production and human health.

The second section addresses citrus pests and disease aspects. Chapter 6 reviews the main citrus diseases caused by viruses and viroids. Chapter 7 examines the quorum quenching mechanism to control citrus bacterial diseases. Chapter 8 discusses citrus-sucking and chewing insect pests.

The last section discusses the use of citrus in human health. Chapter 9 highlights the benefits of citrus consumption on human health. Chapter 10 discusses citrus polyphenols as strategies to overcome multidrug resistance. Finally, Chapter 11 addresses the use of citrus essential oils and citrus terpenes loaded in nanocarriers for treating skin diseases and for transdermal drug release.

Mateus Pereira Gonzatto
Department of Agronomy,
Federal University of Viçosa,
Viçosa, Brazil

Júlia Scherer Santos
Faculty of Pharmacy,
Department of Pharmaceutical Sciences,
Federal University of Juiz de Fora,
Juiz de Fora, Brazil

Section 1

Citrus Physiology and Production Technology

Chapter 1

Introductory Chapter: World Citrus Production and Research

Mateus Pereira Gonzatto and Júlia Scherer Santos

1. Introduction

The set of plants producing citrus fruits is composed of a large number of species from the *Citrus*, *Fortunell*a, and *Poncirus* genera. Among the main species are sweet oranges (*Citrus sinensis*) and bitter oranges (*Citrus aurantium*), mandarins (e.g. *Citrus reticulata*, *Citrus delicia*, *Citrus clementina*, *Citrus unshiu*, and *Citrus nobilis*), lemons (*Citrus limon*), limes (e.g. *Citrus latifoli*a, *Citrus aurantifolia* and *Citrus limettioides*), grapefruits (*Citrus paradisi*), pummelos (*Citrus maxima*), kumquats (*Fortunella* spp.), and trifoliate orange (*Poncirus trifoliata*). Fruits of commercial importance for human consumption are mainly from *Citrus* species. In a small proportion, *Fortunella* species (fruits known as kumquats) are also worthed. Moreover, several hybrids such as tangors (mandarin × sweet orange) and tangelos (mandarin × grapefruit) are also important commercially. On the other hand, *P. trifoliata* and its hybrids *Citrus* spp. are used only as rootstocks, not being edible [1].

2. World production

Among the main fruits produced in the world, citrus are highlighted. The citrus fruits were the second most produced fruit worldwide in 2021, accounting for 161.8 million tons produced in more than 10.2 million hectares. Only bananas and plantains combined exceeded this amount, reaching more than 170.3 million tons [2].

Oranges world production is the most significant within citrus, reaching 75.57 million tons (46.7% of citrus fruit production) in a harvested area of 9.93 million hectares. The second most important fruits are tangerines, with a production of 41.95 million tons (25.9% of citrus fruit production) in a harvested area of 3.11 million hectares. Otherwise, limes and lemons reached 20.83 million tones (12.87% of citrus fruit production) in a harvested area of 1.34 million hectares. Besides, pomelos and grapefruits (*C. paradisi* and *C. máxima*) and other fruits (*Citrus medica, Citrus bergamia, Citrus myrtifolia* and *Fortunella* spp.) are also notorious. Pomelos and grapefruits have a production of 9.56 million tons while other citrus fruits have a production of 13.90 million tons (**Figure 1**) [2].

Among the largest producing countries, China, Brazil, and India can be highlighted. The great productivity of Chine is mainly due to the larger production of tangerines and other citrus fruits. As for Brazil, it is the largest producer of oranges and the largest exporter of orange juice. In turn, India is the world's largest producer of lemons and limes. Additionally, Mexico, Spain, United States of America,

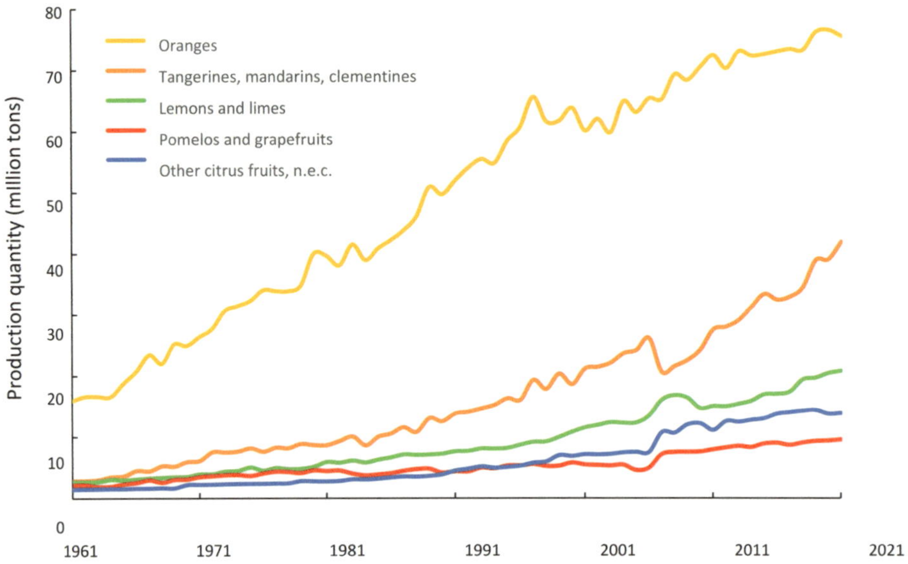

Figure 1.
World production of different citrus fruits between 1961 and 2021 [2].

Türkiye, Egypt, Nigeria, and Iran must also be noted as the top 10 producers of citrus fruits in 2021 [2].

Also, world production of citrus has increased almost 5.5 times in the last 60 years. Notably, in the last 10 years (between 2021 and 2011), there was the highest increase to tangerines production, with more than 1.44 million tons produced per year. In turn, oranges production increased by 312 thousand tones per year while the lemons and limes one increased by 578 thousand tons per year (**Figure 1**) [2].

3. Research on citrus

Scientific production about citrus is equally important in different areas of human knowledge. From the search for the term "citrus" in the "all fields" option in the Scopus and Web of Science databases in the last 5 years (2018–2022), there was an increasing number of publications in both databases [3, 4]. Although the search with the term "all fields" retrieves in very broad number of publications, the Scopus database included a much larger number of publications. In year of 2022, from Scopus database were retrieved 31,103 documents while for Web of Science database a lower number was recovered (2,839 publications) (**Figure 2A**).

Then, a new, more restrictive search was performed in Scopus database, from 2018 to 2022 searching the term "citrus" in the "article title, abstract, and keywords" option. A total of 21,954 results were retrieved, where 66% of these publications are from the last three years (from 2020 to 2022). Therefore, citrus research is a topic of current relevance (**Figure 2A**).

Regarding to the countries that most contributed to publications about citrus in the last 5 years, there are China, United States, India, Brazil, Spain, Italy, Iran, South Korea, Japan, and Pakistan. Of these, six were also the top 10 citrus fruit producers in

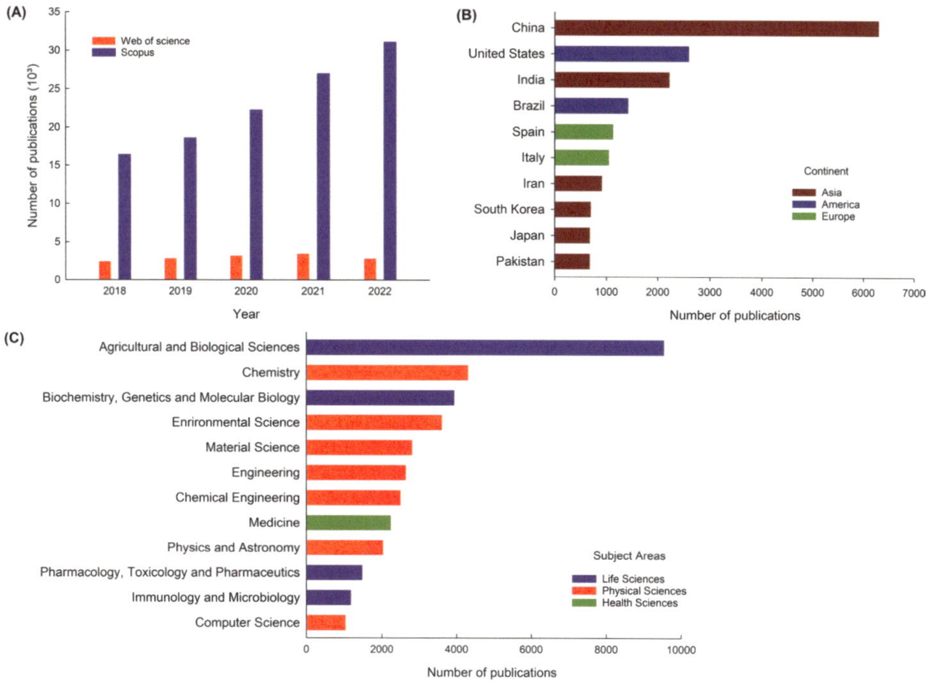

Figure 2.
(A) Number of publications retrieved from Scopus and Web of Science databases searching for the term "citrus" in all fields (ALL(citrus); (B) Most relevant countries when searching for term "citrus" in "article title, abstract and keywords" (TITLE-ABS-KEY(citrus)) in Scopus database; (C) Main knowledge subject areas when term "citrus" was search in article title, abstract and keywords (TITLE-ABS-KEY(citrus)). The searches were performed in the last 5 years (2018 to 2022).

2021, namely: China, Brazil, India, Spain, United States, and Iran [1]. China is the biggest contributor, with 28.66% of the total published. Researchers from USA and India have published more than 2,000 articles, standing for 11.82% and 10.12% of the total articles produced in the period, respectively. Further, with more than 1,000 publications are Brazil, Spain, and Italy. Besides, Iran, South Korea, Japan, and Pakistan had around 600–1,000 publications. Additionally, the great importance of Asian countries to knowledge production related to "citrus" can be noted as they account for more than 50% of total publications (**Figure 2B**) [3].

As to the subject areas by Scopus database classification, 43.4% of the publications were from Agricultural and Biological Sciences. The second most important area was Chemistry (19.6%), followed by Biochemistry, Genetics and Molecular Sciences (17.9%). Also, there was a significant number of publications addressing citrus in other areas of knowledge: Environmental Science, Material Science, Engineering, Chemical Engineering. Yet, some part of publications was applied to the medical and pharmacological area (**Figure 2C**).

Furthermore, many patents are report on citrus. In a search in the Derwent Innovations patent database over the last 5 years (2018–2022) [5], using the term "citrus" as a topic resulted in 6977 patents. The majority of these publications cover the following areas: chemistry, agriculture, biotechnology, instruments, food science, pharmacology. Hence, research on citrus is directed to several areas of knowledge, which shows its importance and the need of continuous increase in its knowledge.

Author details

Mateus Pereira Gonzatto[1*] and Júlia Scherer Santos[2]

1 Department of Agronomy, Federal University of Viçosa, Viçosa, Brazil

2 Faculty of Pharmacy, Department of Pharmaceutical Sciences, Federal University of Juiz de Fora, Juiz de Fora, Brazil

*Address all correspondence to: mateus.gonzatto@ufv.br

References

[1] Talon M, Caruso M, Gmitter FG, editors. The Genus Citrus. Duxford: Elsevier; 2020. p. 538. DOI: 10.1016/C2016-0-02375-6

[2] FAO. Food And Agriculture Organizations of the United Nations [Internet]. 2023. Available from: http://www.fao.org/faostat/en/#data. [Accessed: June 2, 2023]

[3] Scopus. Scopus Database [Internet]. 2023. Available from: http://www.scopus.com. [Accessed: 2023-02-06]

[4] Web of Science. Web of science Database [Internet]. 2023. Available from: www.webofscience.com. [Accessed: June 02, 2023]

[5] Derwent Innovations. Derwent Innovations Index [Internet]. 2023. Available from: https://clarivate.com/products/ip-intelligence/ip-data-and-apis/derwent-world-patents-index. [Accessed: June 02, 2023]

Chapter 2

Abiotic Stresses Management in Citrus

Zeinab Rafie-Rad, Majid Moradkhani, Ahmad Golchin, Taqi Raza and Neal S. Eash

Abstract

Citrus production is affected globally by several environmental stresses. Some citrus-producing regions suffer from severe ecological abiotic stresses, including cold, soil salinity and sodicity, extreme temperature, and drought. These abiotic stresses can alleviate the growth, fruit yield, and quality of citrus. Strategies that attempt to sustain and increase tolerance of citrus against the negative effect of abiotic stresses are the use of antiperspirant compounds, phytohormones, synthetic and natural growth regulators, soil and plant moisture retaining tools and structures, nutrition management, application of organic fertilizers, rootstocks breeding in citriculture, and others. These strategies increase the yield and growth of the plant along with the relative improvement of the fruit quality during the growth and fruiting period, increasing the absorption of water and nutrients, the extensive accumulation of osmolytes and the increase of antioxidant enzymes, changes in the amount of signaling substances, and the expression of genes under stress, increase tolerance to abiotic stresses in citrus fruits. In this review, we tried to provide a summary of the abiotic stress management in citrus by literature.

Keywords: citrus, climate change, soil salinity and sodicity, drought, extreme temperature, cold

1. Introduction

Citrus species are momentous commercial fruit crops globally and are grown in more than 140 countries around the world. Citrus consists of more than 162 species belonging to the order Geraniales, family Rutaceae, and subfamily Aurantoideae. Oranges (*Citrus sinensis* (L.) Osb.), grapefruits (*Citrus paradisi* Macf.), lemons (*Citrus limon* Burm. F.), limes (*Citrus latifolia* Tan. and *Citrus aurantifolia Swingle)*, mandarins (*Citrus reticulata* Blanco), and pummelos (*Citrus maxima* (Burm.) Merr.), are the most common types of citrus fruits that are used as fresh fruit, fruit juice, and concentrate [1, 2]. Citrus is one of the fruit culture parts with the highest production on a global scale (**Figure 1**), and China is the world's largest citrus producer with 44,063,061 tons of production per year. Brazil comes second with 19,652,788 tons of yearly production. With 14,013,000 tons of production per year, India is the

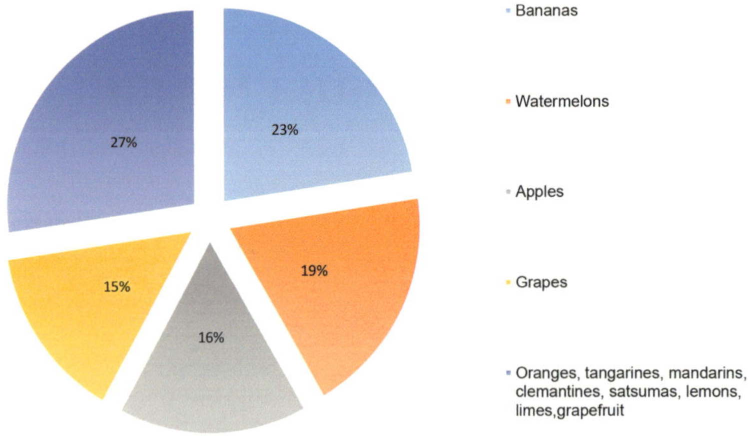

Figure 1.
Global fruit production in 2020, by selected variety (In million metric tons) [1].

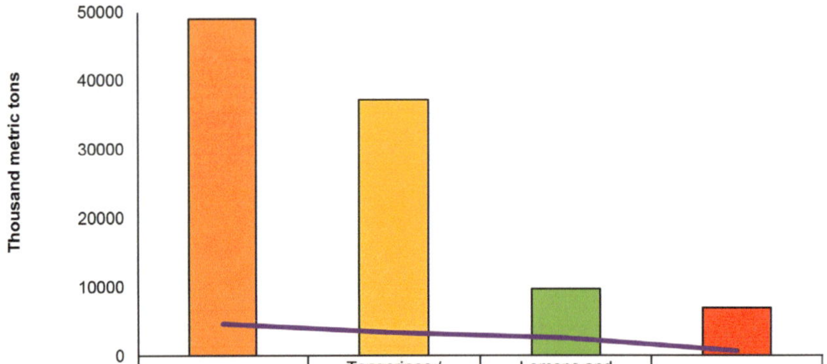

	Oranges	Tangerines / Mandarins	Lemons and Limes	Grapefruit
Production	49014	37211	9749	6963
Export	4637	3347	2541	684

Figure 2.
Global citrus production and export quantity in 2021 [3].

third-largest producer of total citrus. Mexico with 8,756,488 tons, and the United States of America, with 7,230,854 tons of production per year are ranked 4 and 5, respectively [1].

Oranges account for half of the production and represent over 40% of world citrus exports, followed by tangerines/mandarins, lemons/limes, and grapefruit. Global exports are estimated at 11 million tons with oranges representing over 40% and tangerines/mandarins nearly 30%. Exports are propelled by tangerines/mandarins from China, South Africa, and Turkey, and to a lesser extent, higher lemon exports from Mexico, South Africa, and Turkey. South Africa is the largest exporter followed by Turkey and Egypt. U.S. citrus exports are dropping primarily due to lower orange exports. They have not been able to participate in the rising global tangerine or lemon trade due to reduced exportable supplies (**Figure 2**) [3].

Traditional citrus production methods have been successfully used over the years, and global citrus production and export have grown continuously over the past three decades. However, these methods are limited by environmental stresses, and citrus species struggle with many abiotic stresses, including cold, soil salinity and sodicity, extreme temperature, drought, and others [4].

One of the factors that cause extreme temperature, droughts, and cold, caused by severe rains and others, is climate change [5]. Global warming and climate change event have exacerbated the destructive effects of drought stress with various impacts on temperature and rainfall patterns in different areas of the world [6]. In such circumstances, water scarcity and precipitation are considered limiting factors for agriculture and crop productivity in some countries. The minus effects of abiotic stresses generally reduce tree growth, fruit yield, quality, and limit crop productivity. Under normal conditions, citrus trees mostly confront numerous stresses simultaneously, so there is a direct and indirect interplay between approximately all physical abiotic stresses. The morphological, physiological, and biochemical responses of citrus trees exposed to two or more abiotic stress factors can change, depending on stress duration or intensity. Intricate genetic responses to abiotic stresses are polygenic, making them more challenging to detect, control, and manipulate [7].

Soil salinity and drought stress are one of the main factors of yield losses in the world that can reduce the relative water content and leaf water potential. It also prevents cell enlargement more than cell proliferation. As a result, drought and salt stresses reduce plant growth and leaf area, ultimately affecting plants' photosynthesis, respiration, secondary metabolites, and carbohydrate production [4, 8]. As mentioned, climate change causes extreme temperature fluctuations, namely increase or decrease in temperature, and can cause adverse effects on plants in terms of physiology, biochemistry, and gene regulation pathways. Increased temperature causes heat stress in plants, depending on the light intensity, duration, and quality. Also, decreased temperature results in the loss of membrane integrity, leaf damage, electrolyte leakage, impaired photosynthesis, chlorophyll pigments, and protein assembly [9].

As high-temperature stress causes soil dryness and drought stress, soil waterlogging will also be stressful for the plant. Plants grow well by absorbing water through the roots and transpiration through the leaves. If the soil becomes saturated or supersaturated, waterlogging stress occurs. In this case, the leaf pores are closed, chlorophyll is lost, chlorosis is created in the leaves, and as a result, photosynthesis is reduced [10]. Stresses, such as nutrient deficiency and excessiveness, metal toxicity, and ultraviolet irradiance may occur less often, but these stresses decrease the quality and quantity of the products. Reducing the application of organic fertilizers and the imbalanced use of chemical fertilizers are the leading causes of nutritional deficiencies in plants, and we do not explain these stresses here [11]. Therefore, new production programs are essential to obtain crops more tolerant to abiotic stresses. Plant breeding programs, phytohormones, metabolic inhibitors, anti-transpiration and anti-evaporations, and plant nutrition management are critical against abiotic stress. Consequently, in this updated review, we will introduce types of abiotic stresses in citrus fruits and manage the stress tolerance of citrus by providing solutions to deal with these stresses. Hoping to provide readers with good ideas on increasing citrus fruit tolerance under abiotic stress conditions.

2. Introduction of abiotic stresses types in citrus

Citrus species, the most consumed fruit products in the world, are mainly produced in coastal areas in several countries and Mediterranean areas. Production in these areas is affected by abiotic stresses, such as drought, extreme temperature, salinity and sodicity, and others [12]. These stresses cause many reactions in the plant, including the destruction of the photosynthesis system and changes in gene expression and metabolic processes, such as increasing the synthesis of secondary metabolites and production and accumulation of reactive oxygen species (ROS) [13, 14]. It is necessary to entirely understand the reaction of citrus to abiotic stresses at different levels to apply strategies to increase stress tolerance. Plant metabolism is affected by environmental stresses, and studying at such a level is necessary for using stress-reducing compounds. **Figure 3** summarizes the adverse effects of abiotic stresses and some physiological and biochemical responses of plants to these stresses, which we will describe below.

2.1 Drought stress and its management and mitigation strategies

Citrus growth and production are affected by various environmental agents, one of which is drought. Drought, or, in other words, limited access to water is considered one of the most limiting citrus growth and production factors with adverse effects. This stress negatively affects citrus production and will be progressively more intense in some areas due to global climate changes [15, 16]. Climate change event and, consequently, global warming has exacerbated the destructive effects of drought stress with various impacts on temperature and rainfall patterns in different areas [10]. In such circumstances, water scarcity and precipitation consider limiting factors for citrus productivity in some countries, especially in Mediterranean regions.

Drought causes physiological and biochemical changes in the citrus, associated with decreased osmotic potential on the cell surface [17, 18]. Citrus growth and

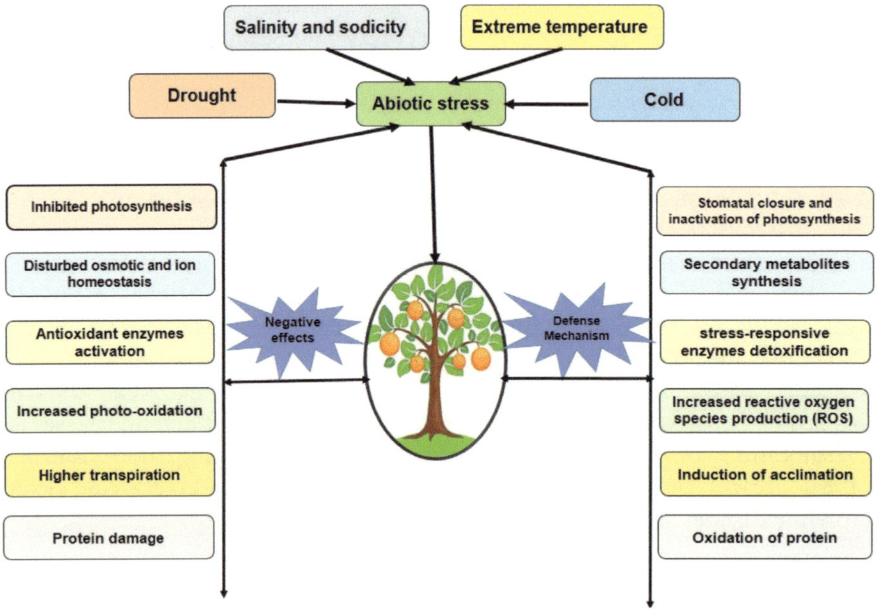

Figure 3.
It summarizes the adverse effects of abiotic stresses and some physiological and biochemical responses of plants to these stresses. Though the consequences of various abiotic stresses are different, the physiological and biochemical responses seem approximately similar. It is noteworthy that the adaptive strategies of plants against abiotic stresses are analogous (the above-summarized information is extracted from the work of [14]).

performance are decreased under drought-stress conditions by changing photosynthesis rates. Also, with the increase in the duration of drought stress, the amount of secondary metabolites changes. In this situation, osmotic pressure-regulating substances, such as proline (Pro), accumulate in the plant by spending much energy and reducing the osmotic potential. Research on lemon [19] and "Nagami" kumquat [20] leaves show a decrease in leaf water potential under drought stress conditions. Another physiological effect of drought stress in citrus is the reduction of leaf chlorophyll. In such conditions, various types of reactive oxygen species (ROS) are produced in photosystem II of photosynthesis. ROS produce compounds, such as malondialdehyde, which cause cell damage. With the accumulation of malondialdehyde, the permeability of the plasma membrane and, thus, ion leakage increases [10, 15]. The more significant influence of fruits, compared to other vegetative organs, in response to drought stress is an exciting object. Drought stress affects the quantity and quality of citrus fruit and causes a decrease in yield because available water is one of the crucial factors for increasing yield [21]. Also, in drought stress conditions, maintaining cell integration through osmotic regulation and then the accumulation of soluble solids in the fruit during the growth period increases the quality properties, such as soluble solids (SS) and titratable acidity (TA). The increase of these quality indicators in drought stress conditions has been reported in Salustiana orange fruit [22].

One of the adverse effects of drought stress in citrus is the fruit-cracking after drought stress, which is evident in citrus cultivars with navels, such as "Navel" and "Valencia" orange (*Citrus sinensis* (L.) Osbeck) and mandarin hybrids (*Citrus reticulata* Blanco). Cracked fruits are prone to decay and lose their ability to be stored. As a result, their economic value decreases. In order to reduce the bursting and the economic loss caused by it, the water required by the plant should be provided in different stages of growth [23]. Executing suitable agricultural practices is generally considered obligatory for overcoming the adverse effects of drought stress [24]. Today, advanced techniques are used in citrus production to increase the soil's water-holding capacity and improve the application of limited water resources. One of the new methods is using superabsorbent polymers (SAP) or hydrophilic polymer gels. These polymers can quickly absorb large amounts of water and gradually provide the water stored in their structure to the plant while drying the environment. In this way, the soil remains wet for a long time without rewatering [25, 26]. Modifying the citrus root system surrounded by these polymers will increase water retention and minerals in the plant growth environment, improve soil texture, increase water penetration and germination, and faster plant growth. Research conducted on Carrizo citrange and Cleopatra mandarin seedlings showed that the application of superabsorbent polymer increased water absorption and leaf chlorophyll content [25]. Superabsorbent polymers provide moisture to the soil and plants and reduce quality indicators of the fruit, such as soluble solids and titratable acidity. The results of research on sweet orange confirm these findings [27]. Another strategy proposed in recent years to deal with the harmful effects of drought stress and the high summer temperature is using antiperspirant and sunlight-reflective compounds, such as kaolin. Kaolin is a natural clay with an aluminum phyllosilicate ($Al_2Si_2O_5(OH)_4$) structure, which is chemically neutral in a wide range of pH changes and does not harm living organisms [28, 29].

High temperature causes drought stress in plants. So, net shading is another technique to improve tree water status and water use efficiency in water deficiency conditions [30, 31]. The most common shade used for this purpose is made of aluminum, polypropylene, polyethylene, and polyester, as well as thin or linen sacks woven together in the form of webs [32, 33]. These nets do not change the natural composition of light; only the light spectrum passing through the net is changed. The benefits of using net shading on plants in drought stress conditions are improving plant water status and yield, delaying fruit maturation to ripen more fruit, and reducing photo-inhibition. In severe water stress conditions, net shade is effective in the long term [34, 35]. Net shading may also harm fruit growth capacity because it reduces the amount of light received by the plant.

Nevertheless, single or multi-layer shades with different light transmission percentages is used [31]. An example of polyethylene shade with layers and different light transmission percentages is presented in **Figure 4**. As mentioned in **Figure 4**, shading was carried out with commercial green and reticular polyethylene shade, which transmits about 70, 50, and 0% of incident light from left to right, respectively.

Other drought stress management methods include exogenous application of plant growth regulators (PGRs), such as abscisic acid (ABA), auxin, gibberellic acid, brassinolide, jasmonates, benzyl-adenine, salicylic acid (SA), and biostimulants. PGRs are nonnutritive organic compounds that control different phases of plant growth and development as secondary stress messengers and play an essential role in reducing these abiotic stress conditions. For example, a study on citrus has shown that jasmonic acid (JA) signaling pathways are effective in water stress. In citrumelo CPB 4475 (*Citrus paradisi* × *Poncirus trifoliate*), a hybrid citrus genotype used as a rootstock, it was found that the exogenous JA application can also effectively diminish the damage caused by drought to plants [37]. Also, rootstock breeding and biotechnological approaches can mitigate climate change effects and increase drought tolerance in citrus [33, 38].

2.2 Extreme temperature stress and its management and mitigation strategies

Climate change has caused severe problems, especially global warming, which has increased the earth's average temperature by four degrees fahrenheit compared to the ice age. **Figure 5** shows the increase in the earth's temperature from 1884 to

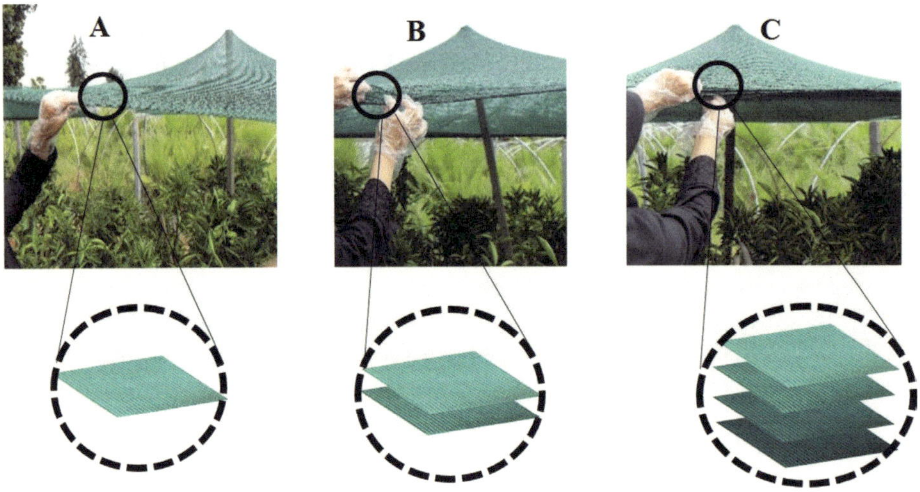

Figure 4.
Green and reticular polyethylene shades with one (A), two (B), and four (C) layers on page mandarin seedlings, from left to right, respectively [36].

2021. Climate change and agriculture are interrelated activities and affect citrus as one of the largest fruit crops in the world. The growth and yield of citrus begin to change when the environmental temperature increases; the water required by the plant decreases [39]. Extreme temperature causes biochemical, morphological, physiological, and genetic changes in citrus. High temperatures in citrus growing regions can lead to significant leaf-to-air vapor pressure difference (D) and excessive leaf temperature in sun-exposed leaves. This high-temperature stress can diminish net CO_2 assimilation, growth, fruit yield, and citrus quality. The critical stages of citrus phenology increase the abscission of reproductive structures and the dropping of young fruit [40]. In citrus, high temperatures cause the generation of reactive oxygen species (ROS), the decline in chlorophyll content, an accumulation of carotenoids, and delayed coloration [41]. Research conducted on Carrizo citrange and Cleopatra mandarin showed that heat increases the phenolic compounds, the composition of secondary metabolites, and the antioxidant capacity of the leaves [42].

The plant's nutritional requirements are higher during the flowering and fruiting of citrus, and trees produce and accumulate the most carbohydrates during the growth stage. Citrus trees have a lot of nutritional requirements during the flowering and fruiting period; therefore, in the growth stage, trees produce and accumulate carbohydrates more than other materials. If nutrient accumulation is insufficient, high temperature may lead to increased respiration and diminishing photosynthesis, which causes nutritional imbalance and thus exacerbates flower and fruit drops [43]. The high temperature increased the fruit drops of Nagami Kumquat [43] and Tosa Buntan pummelo (*Citrus grandis* (L.) Osbeck) trees [44].

Global warming, and, as a result, extreme temperature stress is a dangerous challenge for agricultural products, including citrus fruits, and requires efficient strategies to ensure the production of quality products. One of the effective strategies is kaolin particle film application (**Figure 6**). The kaolin application is a suitable strategy to deal with the increasing temperature of the environment caused by global warming. It can also significantly help to expand the growth of citrus fruits in areas

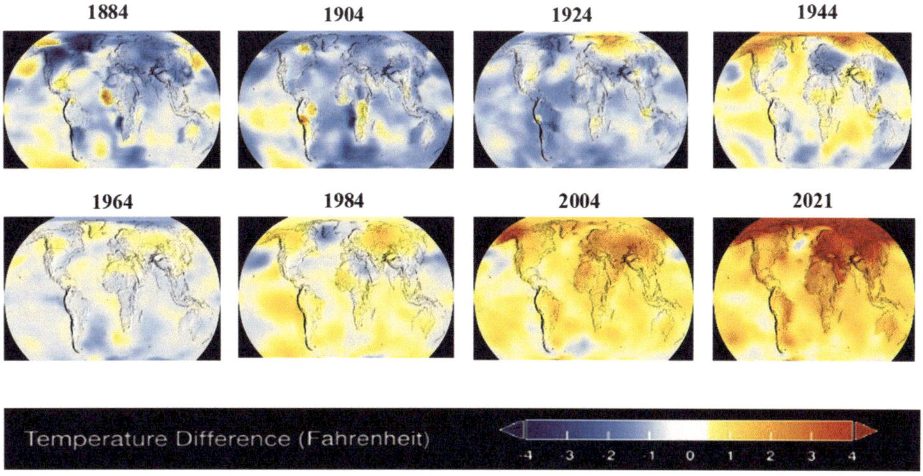

Figure 5.
Global temperature changes from 1884 to 2021 (adapted from NASA/GISS).

Figure 6.
Kaolin clay particles application on page mandarin (Citrus reticulata) trees. The page mandarin trees were treated with 7% (A) and 5% (B) kaolin clay [36].

that suffer from summer heat stress. The main effect of kaolin is to increase solar radiation reflection and reduce, consequently, the temperature of the leaves exposed to light in the hottest hours [29, 34]. By reflecting a part of the light irradiated to the plant, kaolin significantly reduces the temperature of the leaves. It also prevents the accumulation of Pro by increasing the potential and maintaining the relative water content of the leaves in the summer. As a result, it reduces the adverse effects of drought stress. The effects of lowering temperature are reducing water consumption, increasing chlorophyll content, maintaining the quantitative and qualitative characteristics of the product, and preventing bursting and sunburn of citrus fruit, which was found in research conducted on Balady mandarin (*Citrus reticulata* Blanco) [29] and grapefruit [28] have been evident. **Figure 6** shows page mandarin (*C. reticulata*) trees sprayed with kaolin.

In citrus-producing regions, net shading could ameliorate leaf water use efficiency, photosynthesis, and fruit quality, especially in citrus seedlings, where most leaves are exposed to sunlight. The advantages of using these shades include reducing sunlight radiation intensity, especially in hot seasons, soil water maintenance, reducing wind speed, reducing leaf temperature, and photo-inhibition [32, 45]. Growth improvement, increasing fruit yield and quality and net gas exchange, mitigating fruit dropping, and other advantages, such as protecting the plant against bird attacks, insects, rain, and high daily, and night temperatures [46]. Compost application can also reduce soil temperature, allowing better root growth and ultimately decomposition and adding significant organic matter to the soil [30, 47]. Some research were carried out on the effect of methyl jasmonate (MeJA) on citrus under high-temperature stress. Findings found that the JA, and JA-isoleucine accumulation, were induced in heat stress conditions, which can further counteract the harmful effects of extreme temperature stress by closing the stomata and reducing transpiration [21, 48, 49]. In addition, in citrus-growing regions, high temperatures will increase irrigation requirements. Therefore, the pressure on the underground aquifers increases the soil salinity. All these stress factors have undesirable results in citrus cultivation.

2.3 Soil salinity and sodicity stress and its management and mitigation strategies

Soil salinity and sodicity are abiotic stresses that prevent the growth of trees, especially citrus, by limiting water and mineral uptake. Citrus is one of the salt-sensitive and salt-intolerant crops, and their response to soil salinity and sodicity depends on rootstock, scion, soil type, irrigation system, and climate. Changing these factors, under the same irrigation conditions, could generate entirely different results. Salinity reduces growth in citrus trees and causes physiological disorders [50]. Salinity damage happens when the dissolved salt in water is high enough to diminish crop growth. Indirect movement of water in the leaf tissues of citrus can cause accumulation of Cl^- ions affecting transpiration and photosynthesis, and increasing Cl – ions concentrations, accelerates defoliation by enhancing leaf abscission and ethylene production [51]. The most common causes of soil salinity and sodicity are hydrological, geological, and soil processes. Other reasons are incompetent irrigation methods, improper drainage systems, dry weather or insufficient annual rainfall, and remaining salts accumulated in plant root areas [52].

The effect of soil salinity on the citrus growth parameters may be seen as decreased plant height, leaf area, stem diameter, fresh and dry weight, and increased tree senescence [50]. The decrease in growth indices may also be due to the salt osmotic effect on the roots and the toxic ions accumulated in plant organs. The addition of sodium chloride (NaCl) in growth media increments phosphorus, nitrogen, and potassium and alleviates calcium and magnesium in most citrus rootstocks [53]. Citrus physiological responses to salt stress might be attributed to changes in water relations. Citrus species, such as halophyte plants, cannot absorb salts from the soil solution and accumulate them in their tissues to regulate osmotic pressure. Instead, they accumulate secondary metabolites, such as pro and nontoxic inorganic ions, in their cells to overcome salinity stress. Salt accumulation in chloroplasts reduces chlorophyll content and directly affects and reduces photosynthetic activities and yield [50, 54].

Phytophthora root rot in citrus fruit is usually intensified under salinity stress conditions. Also, a reduction in number, weight, and fruit yield is observed in citrus fruit under salt stress. Salinity decreases plant growth and yield and reduces shoot and root biomass [55]. The plant pigment contents decrease in response to salt stress due to CL accumulation, destruction of chlorophyll biosynthesis, and the reduction in iron, magnesium, and manganese in several citrus rootstocks [56]. Reduced chlorophyll content in citrus rootstocks due to NaCl application was also reported [57, 58]. The figure below shows the adverse effects of salinity stress on the nutritional, physiological, and biochemical characteristics of four citrus species (**Figure** 7).

Since citrus fruits are one of the most sensitive plants, salinity stress reduces fruit yield and quality. So, improvement programs should be considered for better productivity of citrus fruits in soils affected by salt. One of the methods of managing salinity stress in citrus fruits is irrigation programming. Irrigation is an essential factor in managing salinity in areas with salinity stress. Increasing irrigation periodicity is recommended to leach the salts and minimize the salt concentration in the root zone [57]. Leaf and trunk damage related to the absorption of salt can be reduced by using micro-irrigation systems. Another strategy to diminish salinity stress is recommended repetitious fertilization or spreading dry fertilizers through fertigation. It is necessary to mention that nutrient fertilizers should not contain chloride (CL) or sodium (Na). In areas where the soil is sodium, calcium source application (gypsum; $CaSO_4$)

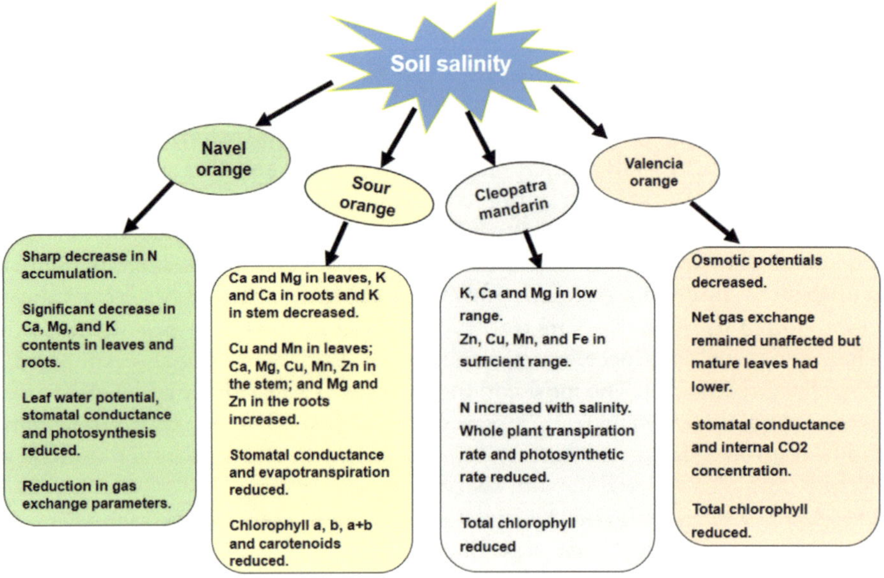

Figure 7.
Effect of soil salinity on nutritional, physiological, and biochemical parameters in four citrus species [59–61].

reduces the adverse effect of sodium on shoot growth and improves plant growth in these conditions [50, 62].

Plant breeding and genetic manipulations are other ways of managing salinity and sodicity stress conditions, which increase the plant's adaptation to saline environments and salt tolerance. Currently, inorganic and organic conditioners like organic residues, phosphor-gypsum, and H_2SO_4, calcium application, drainage management, the genetic use of halophytic traits, arbuscular mycorrhizal, and avoiding cultivation of lands with high groundwater are being used to progress citrus productivities in salt-affected soils [62–65]. Quantitative trait locus (*QTL*) mapping is one way to identify salinity tolerance genes. This way has focused on characterizing genes encoding proteins involved in decreasing the amount of Na^+ from the root to the shoot [66].

Since drought stress is the beginning of creating salinity stress in the soils of dry and salty areas, therefore, methods, such as the use of superabsorbent, netting shades, the application of phytohormones, such as abscisic acid (ABA), polyamines, and chemical priming, will improve the performance of citrus fruits under salinity stress condition [34]. The application of nitrate and other compounds derived from nitrogen (urea or ammonium) has positively affected citrus morpho-physiological and biochemical responses under salt stress conditions [67]. Nitrate appears to stimulate photosynthesis and growth parameters as well as reduce leaf drop. Also, the increased nitrogen in leaf biomass leads to chloride dilution [68]. The beneficial effect of Paclobutrazol (PBZ) under salinity on the accumulation of photosynthetic pigments, phytohormones, and root morphology, such as size, number of lateral roots, and dry weight of roots, has also been documented [69].

2.4 Cold stress and its management and mitigation strategies

Low temperature is the main restrictive factor for citrus growth and productivity worldwide. Citrus is grown at temperatures between 12.8 and 38°C; low temperatures below 13°C limit vegetative growth and fruit growth and delay maturity. The threshold temperature that destroys shoots is −12°C, but some citrus fruit can tolerate a temperature of −10°C [70]. *Poncirus trifoliata* (L.) Raf. is the most tolerant citrus rootstock to low temperatures [71]. The low ambient temperature that causes cold stress is a principal environmental abiotic stress. In these conditions, the production of reactive oxygen species (ROS), such as superoxide radicals, hydrogen peroxide, hydroxyl, and singlet oxygen is increased, which leads to ion leakage and water soaking **Figure 8**, and finally, leaves are destroyed. ROS produced in the chloroplast can destroy cellular components, such as proteins, pigments, membranes, lipids, and nucleic acids [39, 73].

Additionally, mechanical methods can decrease these damages to increase plant resistance against cold and frost damage. One of the strategies is the osmotic balance reaction to maintain plant water content. These activities are affected by osmotic pressure regulatory concentration compounds, such as secondary metabolites, including proline, and some inorganic ions, such as potassium. Potassium increases cells' tolerance against cold stress because it affects the freezing point of the liquid inside the vacuoles [74]. High potassium levels protected cells against freezing by lowering the freezing point of the cell solution [75]. Potassium can affect plant survival under different environmental stresses by creating osmotic balance and protection against oxidative damage. Potassium causes leakage and maintenance of water in plant tissues with increasing osmotic pressure and cell membrane fluidity; thus, it prevents cell membrane rupture and plant tissue damage and increases electrolyte leakage [76].

Potassium substantially affects stomata movement and water relation (turgor regulation and osmotic adjustment) in plants under cold conditions. The present results agree with those reported on Sour orange (*C. aurantium* L.) seedlings [77]. Numerous studies indicated that applying abscisic acid (ABA) may increase leaf water content under low-temperature stress [78]. The grafting technique and the use of tolerant rootstocks to cold stresses allow farmers to protect crops against various abiotic stresses, especially cold stress. A study in common clementine with a tetraploid Carrizo Citrange rootstock showed enhanced natural chilling stress tolerance [79].

Figure 8.
Damaged leaves of sour orange (Citrus aurantium L.) seedlings were exposed to 0, −3, and − 6°C for 24 hours [72].

3. Conclusions

Citrus species is the world's most productive and widely consumed horticultural crops. The factors limiting citrus growth in tropical and subtropical climates are significantly different. Due to abiotic stresses, such as drought, salinity, and high and low temperature, citrus production confronts risks. For this reason, it is crucial using strategies to manage these stresses. Drought and extreme temperatures caused heavy fruit drops and a decline in yield, and it also increases the cracking and folding of fruit and reduces the product yield. Therefore, measures to prevent these problems should be taken in case of severe drought and high temperatures. To manage drought stress and high temperature, using superabsorbents, kaolin, and pure shade is less expensive than producing the product. Also, drought stress management and high-temperature in hot and dry areas reduce the possibility of salinity stress.

Kaolin particle film and shading net reduce water loss through the increasing reflectance of ultraviolet and infrared radiations, thereby reducing leaves and fruit tissue temperature. Kaolin application could be considered an implement to be used in tropical regions to improve plant acclimation to extreme temperatures and high radiation levels in citrus. Maintaining soil moisture, reducing water consumption, and taking advantage of suitable yield along with relative improvement of fruit quality of growth and fruiting period of citrus is essential. For this purpose, soil amendment with superabsorbent polymers to provide moisture in the root area and increase available water in water shortage conditions improve some physiological, biochemical, and phytochemical properties of citrus. Also, the uniformity of humidity during the plant growth period with the use of superabsorbent plays a role of importance in reducing other indicators, such as fruit bursting. Therefore, it is recommended to use this superabsorbent polymer in areas that face water shortages and improper distribution of precipitation. Exogenous application of phytohormones and hormones and even suitable nutrition of citrus fruits can also increase plant tolerance against abiotic stresses. Among the methods that can protect the plant from low-temperature stress are grafting techniques, plant breeding programs, osmotic balance reactions to maintain plant water content secondary metabolites, proline, and inorganic ions, such as potassium.

Citrus is salt-sensitive, and soil salinity reduces citrus growth and causes physiological disorders. Salt stress lowers stomatal conductance, net CO_2 assimilation, and water potential of citrus leaves. Additionally, it causes an excessive concentration of sodium or chloride in citrus leaves. Increasing irrigation periodicity, fertilization and fertigation, Plant breeding, and genetic manipulations are among the methods of managing salinity stress in citrus fruits. In the presence of an adequate concentration of calcium in saline irrigation water, calcium improved the effects of saline on the plant's growth. Therefore, the plants could tolerate the effects of high salinity concentration. The authors' practical experience has shown that abiotic stresses occur worldwide in almost all citrus-growing areas. This research has attempted to explain citrus fruit management and improvement strategies to withstand abiotic stresses in citrus fruit production.

Conflict of interest

The authors have no conflict of interest with any person or institution.

Author details

Zeinab Rafie-Rad[1*], Majid Moradkhani[2], Ahmad Golchin[1], Taqi Raza[3] and Neal S. Eash[3]

1 Faculty of Agriculture, Department of Soil Science, University of Zanjan, Zanjan, Iran

2 Faculty of Agricultural Sciences, Department of Agronomy, Takestan Branch, Islamic Azad University, Takestan, Iran

3 Department of Biosystems Engineering and Soil Science, Institute of Agriculture, University of Tennessee, Knoxville, USA

*Address all correspondence to: z_rafierad83@yahoo.com

References

[1] FAO. Citrus Fruit Statistical Compendium. FAO; 2021

[2] Cimen B, Yesiloglu T. Rootstock Breeding for Abiotic Stress Tolerance in Citrus. London, UK.: IntechOpen; 2016

[3] Foreign Agricultural Service/USDA. Citrus: World Markets and Trade. Florida, USA: Foreign Agricultural Service/USDA; 2022

[4] Mall M, Kumar R, Akhtar MQ. Chapter 1- horticultural crops and abiotic stress challenges. In: Chandra Rai A, Rai A, Kumar Rai K, Rai VP, Kumar A, editors. Stress Tolerance in Horticultural Crops. Sawston, Cambridge, UK: Woodhead Publishing; 2021. pp. 1-19

[5] Hasan N, Kamruzzaman M, Islam S, Hoque H, Bhuiyan FH, Prodhan SH. Development of partial abiotic stress tolerant Citrus reticulata Blanco and Citrus sinensis (L.) Osbeck through agrobacterium-mediated transformation method. Journal of Genetic Engineering and Biotechnology. 2019;**17**(1):14

[6] Trouet V, Babst F, Meko M. Recent enhanced high-summer North Atlantic jet variability emerges from three-century context. Nature Communications. 2018;**9**(1):180

[7] Boari F, Donadio A, Schiattone MI, Cantore V. Particle film technology: A supplemental tool to save water. Agricultural Water Management. 2015;**147**:154-162

[8] Syvertsen JP, Garcia-Sanchez F. Multiple abiotic stresses occurring with salinity stress in citrus. Environmental and Experimental Botany. 2014;**103**:128-137

[9] Forni C, Duca D, Glick BR. Mechanisms of plant response to salt and drought stress and their alteration by rhizobacteria. Plant and Soil. 2017;**410**(1):335-356

[10] Mahajan S, Tuteja N. Cold, salinity and drought stresses: An overview. Archives of Biochemistry and Biophysics. 2005;**444**(2):139-158

[11] Pan J, Sharif R, Xu X, Chen X. Mechanisms of waterlogging tolerance in plants: Research Progress and prospects. Frontiers in Plant Science. 2021;**11**:627331

[12] Syvertsen JP. Aspects of Stress Physiology of Citrus. Leuven: Belgium: International Society for Horticultural Science (ISHS); 2017

[13] Han S, Tang N, Jiang H-X, Yang L-T, Li Y, Chen L-S. CO2 assimilation, photosystem II photochemistry, carbohydrate metabolism and antioxidant system of citrus leaves in response to boron stress. Plant Science. 2009;**176**(1):143-153

[14] Meena KK, Sorty AM, Bitla UM, Choudhary K, Gupta P, Pareek A, et al. Abiotic stress responses and microbe-mediated mitigation in plants: The omics strategies. Frontiers in plant science. 2017;**8**:172

[15] Ziogas V, Tanou G, Morianou G, Kourgialas N. Drought and salinity in Citriculture: Optimal practices to alleviate salinity and water stress. Agronomy. 2021;**11**(7):1283

[16] Karimi V, Karami E, Keshavarz M. Climate change and agriculture: Impacts and adaptive responses in Iran. Journal of Integrative Agriculture. 2018;**17**(1):1-15

[17] Evans JP. 21st century climate change in the Middle East. Climatic Change. 2009;**92**(3):417-432

[18] García-Sánchez F, Syvertsen JP, Gimeno V, Botía P, Perez-Perez JG. Responses to flooding and drought stress by two citrus rootstock seedlings with different water-use efficiency. Physiologia Plantarum. 2007;**130**(4):532-542

[19] Ortuño MF, Alarcón JJ, Nicolás E, Torrecillas A. Interpreting trunk diameter changes in young lemon trees under deficit irrigation. Plant Science. 2004;**167**(2):275-280

[20] Chang Y-C, Chang Y-S, Lin L-H. Response of shoot growth, photosynthetic capacity, flowering, and fruiting of potted 'Nagami' kumquat to different regulated deficit irrigation. Horticulture, Environment, and Biotechnology. 2015;**56**(4):444-454

[21] Zandalinas SI, Balfagón D, Arbona V, Gómez-Cadenas A. Regulation of citrus responses to the combined action of drought and high temperatures depends on the severity of water deprivation. Physiologia Plantarum. 2018;**162**(4):427-438

[22] García-TejeroI,Jiménez-BocanegraJA, Martínez G, Romero R, Durán-Zuazo VH, Muriel-Fernández JL. Positive impact of regulated deficit irrigation on yield and fruit quality in a commercial citrus orchard [Citrus sinensis (L.) Osbeck, cv. Salustiano]. Agricultural Water Management. 2010;**97**(5):614-622

[23] Li J, Chen J. Citrus fruit-cracking: Causes and occurrence. Horticultural Plant Journal. 2017;**3**(6):255-260

[24] Cai X, Zeng R, Kang Won H, Song J, Valocchi AJ. Strategic planning for drought mitigation under climate change. Journal of Water Resources Planning and Management. 2015;**141**(9):04015004

[25] Arbona V, Iglesias DJ, Jacas J, Primo-Millo E, Talon M, Gómez-Cadenas A. Hydrogel substrate amendment alleviates drought effects on young citrus plants. Plant and Soil. 2005;**270**(1):73-82

[26] Cea D, Bonomelli C, Mártiz J, Gil PM. Response of potted citrus trees subjected to water deficit irrigation with the application of superabsorbent polyacrylamide polymers. Agronomy. 2022;**12**(7):1546

[27] Pérez-Pérez JG, Romero P, Navarro JM, Botía P. Response of sweet orange cv 'lane late' to deficit-irrigation strategy in two rootstocks. II: Flowering, fruit growth, yield and fruit quality. Irrigation Science. 2008;**26**(6):519

[28] Jifon JL, Syvertsen JP. Kaolin particle film applications can increase photosynthesis and water use efficiency of 'ruby Red' grapefruit leaves. Journal of the American Society for Horticultural Science jashs. 2003;**128**(1):107-112

[29] Ennab HA, El-Sayed SA, El-Enin MMSA. Effect of kaolin applications on fruit sunburn, yield and fruit quality of Balady mandarin (Citrus reticulata, Blanco). Menoufia Journal of Plant Production. 2017;**2**(2):129-138

[30] Simon-Grao S, Simon I, Lidon V, Conesa A, Manera J, Brotons JM, et al. Effects of shade screens and mulching on the color change of fruits from "Fino 49" lemon trees irrigated with water of different salinity or irrigation regimes. Scientia Horticulturae. 2016;**209**:316-322

[31] Abouatallah A, Salghi R, El Fadl A, Hammouti B, Zarrouk A. Shading nets usefulness for water saving on citrus orchards under different irrigation doses. Current World Environment. 2012;**7**(1):13

[32] Jifon JL, Syvertsen JP. Moderate shade can increase net gas exchange and

reduce photoinhibition in citrus leaves†. Tree Physiology. 2003;**23**(2):119-127

[33] Eaa M, Refaie KM, Mohamed AAA, Hashem FA. Water budget economy of navel Orange production under screen net. International Journal of Pure Agricultural Advances. 2017;**1**(1):10-23

[34] Gimeno V, Díaz-López L, Simón-Grao S, Martínez V, Martínez-Nicolás JJ, García-Sánchez F. Foliar potassium nitrate application improves the tolerance of Citrus macrophylla L. seedlings to drought conditions. Plant Physiology and Biochemistry. 2014;**83**:308-315

[35] Raveh E, Cohen S, Raz T, Yakir D, Grava A, Goldschmidt EE. Increased growth of young citrus trees under reduced radiation load in a semi-arid climate 1. Journal of Experimental Botany. 2003;**54**(381):365-373

[36] Rafie-Rad Z. The Effects of Shading, Soil Application of Hydrogel and Foliar Application of Salicylic Acid and Kaolin on Growth Parameters of Page Mandarin (*Citrus reticulata)* (v. Page) under Drought Stress. Iran: Zanjan University; 2020

[37] de Ollas C, Hernando B, Arbona V, Gómez-Cadenas A. Jasmonic acid transient accumulation is needed for abscisic acid increase in citrus roots under drought stress conditions. Physiologia Plantarum. 2013;**147**(3):296-306

[38] Shafqat W, Naqvi SA, Maqbool R, Haider MS, Jaskani MJ, Khan IA. Climate Change and Citrus. London, UK.: IntechOpen; 2021

[39] Kim M, Kang SB, Yun SK, Kim SS, Joa J, Park Y. Influence of excessively high temperatures on the fruit growth and physicochemical properties of Shiranuhi mandarin in plastic-film greenhouse cultivation. Plants. 2021;**10**(8):1525

[40] Kimura Y, Naeshiro M, Tominaga Y, Anai T, Komai F. Metabolite composition of grapefruit (*Citrus paradisi*) grown in Japan depends on the growing environment and harvest period. The Horticulture Journal. 2017;**86**(4):543-551

[41] Ribeiro RV, Machado EC, Santos MG, Oliveira RF. Seasonal and diurnal changes in photosynthetic limitation of young sweet orange trees. Environmental and Experimental Botany. 2009;**66**(2):203-211

[42] Zandalinas SI, Balfagón D, Arbona V, Gómez-Cadenas A. Modulation of antioxidant defense system is associated with combined drought and heat stress tolerance in citrus. Frontiers in Plant Science. 2017;**8**:953

[43] Chang Y-C, Lin T-C. Temperature effects on fruit development and quality performance of Nagami kumquat (*Fortunella margarita* [Lour.] Swingle). The Horticulture Journal. 2020;**89**(4):351-358

[44] Susanto S, Nakajima Y, Hasegawa K. Effect of different day temperatures on flowering and fruiting in Tosa Buntan Pummelo (*Citrus grandis* (L.) Osbeck). Environment Control in Biology. 1991;**29**(3):97-105

[45] Otero A, Goni C, Jifon JL, Syvertsen JP. High Temperature Effects on Citrus Orange Leaf gas Exchange, Flowering, Fruit Quality and Yield. Leuven: Belgium: International Society for Horticultural Science (ISHS); 2011

[46] Lopez G, Boini A, Manfrini L, Torres-Ruiz JM, Pierpaoli E, Zibordi M, et al. Effect of shading and water stress on light interception, physiology and

yield of apple trees. Agricultural Water Management. 2018;**210**:140-148

[47] Kourgialas NN, Anyfanti I, Karatzas GP, Dokou Z. An integrated method for assessing drought prone areas-water efficiency practices for a climate resilient Mediterranean agriculture. Science of the Total Environment. 2018;**625**:1290-1300

[48] Xiong B, Wang Y, Zhang Y, Ma M, Gao Y, Zhou Z, et al. Alleviation of drought stress and the physiological mechanisms in citrus cultivar (Huangguogan) treated with methyl jasmonate. Bioscience, Biotechnology, and Biochemistry. 2020;**84**(9):1958-1965

[49] Balfagón D, Zandalinas SA-O, Gómez-Cadenas AA-O. High temperatures change the perspective: Integrating hormonal responses in citrus plants under co-occurring abiotic stress conditions. Physiologia Plantarum. 2019;**165**(2):183-197

[50] Alam A, Ullah H, Attia A, Datta A. Effects of salinity stress on growth, mineral nutrient accumulation and biochemical parameters of seedlings of three citrus rootstocks. International Journal of Fruit Science. 2020;**20**(4):786-804

[51] Arbona V, López-Climent MF, Mahouachi J, Pérez-Clemente RM, Abrams SR, Gómez-Cadenas A. Use of persistent analogs of Abscisic acid as palliatives against salt-stress induced damage in citrus plants. Journal of Plant Growth Regulation. 2006;**25**(1):1-9

[52] Hussain S, Morillon R, Anjum MA, Ollitrault P, Costantino G, Luro F. Genetic diversity revealed by physiological behavior of citrus genotypes subjected to salt stress. Acta Physiologiae Plantarum. 2015;**37**(1):1-10

[53] Zhang M, Liu Y, Han G, Zhang Y, Wang B, Chen M. Salt tolerance mechanisms in trees: Research progress. Trees. 2021;**35**(3):717-730

[54] Simpson CR, Nelson SD, Melgar JC, Jifon J, King SR, Schuster G, et al. Growth response of grafted and ungrafted citrus trees to saline irrigation. Scientia Horticulturae. 2014;**169**:199-205

[55] Raga V, Intrigliolo DS, Bernet GP, Carbonell EA, Asins MJ. Genetic analysis of salt tolerance in a progeny derived from the citrus rootstocks Cleopatra mandarin and trifoliate orange. Tree Genetics & Genomes. 2016;**12**(3):34

[56] Acosta-Motos JR, Ortuño MF, Bernal-Vicente A, Diaz-Vivancos P, Sanchez-Blanco MJ, Hernandez JA. Plant responses to salt stress: Adaptive mechanisms. Agronomy. 2017;**7**(1):18

[57] Khoshbakht D, Ramin AA, Baninasab B. Effects of sodium chloride stress on gas exchange, chlorophyll content and nutrient concentrations of nine citrus rootstocks. Photosynthetica. 2015;**53**(2):241-249

[58] Balal RM, Ashraf MY, Khan MM, Jaskani MJ, Ashfaq M. Influence of salt stress on growth and biochemical parameters of citrus rootstocks. Pakistan Journal of Botany. 2011;**43**(4):2135-2141

[59] Tozlu I, Guy CL, Moore GA. Tolerance Mechanisms to Salinity Stress in Citrus and Poncirus. Leuven: Belgium: International Society for Horticultural Science (ISHS); 2002

[60] García-SÁnchez F, Martínez V, Jifon J, Syvertsen JP, Grosser JW. Salinity reduces growth, gas exchange, chlorophyll and nutrient concentrations in diploid sour orange and related allotetraploid somatic hybrids. The

Journal of Horticultural Science and Biotechnology. 2002;**77**(4):379-386

[61] Zekri M. Salinity and calcium effects on emergence, growth and mineral composition of seedlings of eight citrus rootstocks. Journal of Horticultural Science. 1993;**68**(1):53-62

[62] Daba AW, Qureshi AS. Review of soil salinity and Sodicity challenges to crop production in the lowland irrigated areas of Ethiopia and its management strategies. Land. 2021;**10**(12):1377

[63] Kitila K, Chala A, Workina M. Effect of gypsum and compost application in reclaiming sodic soils at small scale irrigation farm in Bora District of east Shewa zone, Oromia. Ethiopia. Agriways. 2020;**8**:28-44

[64] Mukhopadhyay R, Sarkar B, Jat HS, Sharma PC, Bolan NS. Soil salinity under climate change: Challenges for sustainable agriculture and food security. Journal of Environmental Management. 2021;**280**:111736

[65] Khalil HA, Eissa AM, El-Shazly SM, Aboul Nasr AM. Improved growth of salinity-stressed citrus after inoculation with mycorrhizal fungi. Scientia Horticulturae. 2011;**130**(3):624-632

[66] Tilbrook J, Roy S. Salinity tolerance. In: Plant Abiotic Stress. Santa Barbara, CA: Wiley Online Library; 2013. pp. 133-178

[67] Álvarez-Aragón R, Rodríguez-Navarro A. Nitrate-dependent shoot sodium accumulation and osmotic functions of sodium in Arabidopsis under saline conditions. The Plant Journal. 2017;**91**(2):208-219

[68] Khoshbakht D, Ghorbani A, Baninasab B, Naseri LA, Mirzaei M. Effects of supplementary potassium nitrate on growth and gas-exchange characteristics of salt-stressed citrus seedlings. Photosynthetica. 2014;**52**(4):589-596

[69] Dubey AK, Srivastav M, Singh AK, Pandey RN. Growth and physiological response of salt-sensitive and salt-tolerant rootstocks of citrus to paclobutrazol under salt stress. The Indian Journal of Agricultural Sciences. 2011;**79**(8):595-599

[70] Nesbitt ML, Ebel RC, Findley D, Wilkins B, Woods F, Himelrick D. Assays to assess freeze injury of Satsuma mandarin. Hort Science Hort Sci. 2002;**37**(6):871-877

[71] Wang M, Zhang X, Liu J-H. Deep sequencing-based characterization of transcriptome of trifoliate orange (Poncirus trifoliata (L.) Raf.) in response to cold stress. BMC Genomics. 2015;**16**(1):555

[72] Rafie-Rad Z. The Effect of Potassium on some Physiological, Morphological and Bio Chem Ical Indices in *Citrus aurantium* Seedling under Low-Temperature Stress. Iran: Gorgan University of Agricultural Sciences and Natural Resources; 2014

[73] Primo-Capella A, Martínez-Cuenca M-R, Forner-Giner MÁ. Cold stress in citrus: A molecular, physiological and biochemical perspective. Horticulturae. 2021;**7**(10):340

[74] Obreza TA. Importance of potassium in a Florida citrus nutrition program. Better crops. 2003;**87**(1):19-22

[75] Kant S, Kafkafi U, Pasricha N, Bansal S. Potassium and Abiotic Stresses in Plants. Vol. 233. Gurgaon: Potassium for sustainable crop production Potash Institute of India; 2002. p. 251

[76] Yelenosky G, Guy CL. Freezing tolerance of citrus, spinach, and petunia

leaf tissue 1: Osmotic adjustment and sensitivity to freeze induced cellular dehydration. Plant Physiology. 1989;**89**(2):444-451

[77] Rafie Rad Z, Dordipour E, Tajvar Y. Evaluation of low temperature tolerance indices in seedlings of Citrus aurantium under potassium nitrate nutrition. Iranian Journal of Plant Physiology. 2017;**7**(2):2007-2015

[78] Xue-Xuan X, Hong-Bo S, Yuan-Yuan M, Gang X, Jun-Na S, Dong-Gang G, et al. Biotechnological implications from abscisic acid (ABA) roles in cold stress and leaf senescence as an important signal for improving plant sustainable survival under abiotic-stressed conditions. Critical Reviews in Biotechnology. 2010;**30**(3):222-230

[79] Oustric J, Morillon R, Luro F, Herbette S, Lourkisti R, Giannettini J, et al. Tetraploid Carrizo citrange rootstock (Citrus sinensis Osb.×Poncirus trifoliata L. Raf.) enhances natural chilling stress tolerance of common clementine (Citrus Clementina Hort. Ex tan). Journal of Plant Physiology. 2017;**214**:108-115

Chapter 3

Citrus Polyembryony

Angel Villegas-Monter, Elisa Del Carmen Matínez-Ochoa, María Andrade-Rodríguez and Itzel Villegas-Velázquez

Abstract

Polyembryony is a type of sporophytic apomixis common in citrus species. Previous studies discovered that embryo traits relate to their sexual or asexual origin. Smaller embryos at the micropyle end are considered zygotic embryos, whereas larger embryos are nucellar. Early embryogenesis in the ovule of apomictic citrus promotes the development of nucellar embryos. The chalaza region inhibits the early development of the zygotic and nucellar embryos; thus, both embryos must grow at the micropyle end. Numerous researchers agree that highly polyembryonic cultivars produce nucellar seedlings more often as the zygotic embryos cannot survive field conditions. Thus, the selection of polyembryonic genotypes facilitates clonal propagation. This chapter analyzes the factors that affect polyembryony in citrus.

Keywords: apomixis, rootstocks and cultivars, nucellar plants

1. Introduction

Angiosperms have two reproduction routes: gametophytic (sexual) and apomictic (asexual). The first form of reproduction leads to seeds with an embryo from the union of the egg cell nucleus with one of the generative nuclei from the male gamete producing a plant with genetic characteristics different from the female parent. This process promotes genetic diversity through hybridization and the adaptation of plants that allows the conservation of species [1]. We will start this chapter by describing the apomictic process and the evolutionary mechanism of alternate pathways that have allowed cloning plants using seeds [2], including most citrus species.

Various mechanisms intervene to generate asexual embryos, eluding the fundamental aspects of sexual reproduction: meiosis and fertilization [3]. According to the mechanism of embryogenesis, apomixis comprises gametophytic apomixis (apospory and diplospory) and adventitious embryony. In these, apomictic embryos have maternal inheritance, but each mechanism is associated with a different probability of producing sexual offspring, varied selection pressure to maintain male fertility, and expected levels of genetic diversity within populations [4]. In apospory, the initial cells form unreduced embryo sacs from direct mitosis and can coincide with sexual embryogenesis: if endosperm forms, the process is pseudogamic, or autonomous if it does not [5]. This second route, which occurs in less than 1% of angiosperms, comprises the development of embryos from somatic cells.

Similarly, in diplospory, the megaspore does not undergo meiosis and through mitosis forms an unreduced embryo sac with cells arranged as in the Polygonum type (sexual reproduction); but the sexual process is wholly compromised [5]. Whereas in adventitial apomixis, the development of the sexual embryo sac seems to develop normally, and after fertilization and development of the sexual embryo, somatic embryos develop from nucellar or integumental cells [6, 7]. In citrus, adventitious embryonic development does not occur without zygote formation (apomixis is facultative), depends on endosperm formation, and is simultaneous with the development of sexual embryos [6, 8, 9]. Therefore, when the embryo sac expands, the embryogenic

Citrus genotypes	Polyembryony (%)	Range of embryos/seed	Author
Amblicarpa mandarin	82.1	2-15	[11]
Amblicarpa mandarin	91.5	4.9	[12]
C-35 citrange	86.5	3.95	[12]
Cipó sweet orange	98	28.8	[13]
Cleopatra mandarin	84.7	1-26	[11]
Cleopatra mandarin	97.1	8.01	[14]
Cravo Santa Cruz Rangpur lime	58.5	1.97 ± 1.08	[15]
Flying dragon trifoliate orange	40	1.70 ± 1.11	[15]
Indio citradarin	100	13.52 ± 4.40	[15]
Minneola tangelo	77.5	3.45	[12]
Morton citrange	86.8	5.87	[14]
Rangpur lime	43.3	1.87	[14]
Riverside citradarin	100	12.97 ± 3.86	[15]
Rough lemon	96.2	4.89	[14]
Rusk citrange	92.3	5.05	[14]
San Diego citradarin	97.5	6.29 ± 3.19	[15]
Smooth Flat Seville sour orange	29	22.7	[13]
Sunki mandarin	66.2	1.8	[14]
Sunki Tropical mandarin	100	8.97 ± 2.76	[15]
Swingle citrumelo	64	2.48 ± 1.47	[15]
Swingle citrumelo	65	2.96	[14]
Troyer citrange	95.1	6.95	[14]
Valencia orange	95.5	4.5	[12]
Volkamerian lemon	37.8	2.50	[11]
Volkamerian lemon	85.5	3.15	[12]
Volkamerian lemon	52.2	2.57	[14]
Yuma citrange	21	1.36	[14]
Yuma citrange	31	4	[13]

Table 1.
Percentage of polyembryony and embryos per seed in apomictic genotypes of citrus.

cells of the nucellus enter the endosperm, competing for space and nutrients with the zygotic embryos that may or may not complete their development [10].

Adventitial embryony or sporophytic apomixis frequently occurs in Rutaceae, Celastraceae, and Orchidaceae; furthermore, it is common in tropical and subtropical trees and shrubs [4]. Polyembryony in citrus is a type of sporophytic apomixis that occurs in most of its species and hybrids (**Table 1**). For example, in 12 genotypes (Sikkim, Thorny, Kinnow and Cleopatra mandarin, Calamondin, Nova×Orlando hybrid, Minneola tangelo, Thornton, Parramatta, Parson Brown, Smooth Flat Seville, and rough lemon rootstock) rates of polyembryonic seeds ranged from 69.8% to 91.4%, with up to 14 embryos per seed [8]. The following genotypes do not show polyembryony, thus classified as non-apomictic: *Citrus hongheensis*, *C. macropera*, *C. medica*, *C. maxima*, *C. clementina*, *Chirita mangshanensis*, and some mandarin hybrids [16, 17]. Although polyembryony is not a common biological phenomenon among angiosperms, it is a form of reproduction that has been studied for 14 decades. According to Cook [18], the oldest record of polyembryony in orange seeds (*Citrus sinensis* L. Osbeck) occurred in 1719 by Leeuwenhoek. Later, Strasburger [19] worked with different species and established that in sour orange (*Citrus aurantium*), all the embryos not derived by the cross originated from nucellar cells, and he called them "adventitious embryos." Since then, polyembryony in citrus has become a form of vegetative propagation, with the problem that for genetic improvement, seedlings from hybrid embryos cannot be differentiated from seedlings from adventitious embryos. Then, hybrid embryos can be differentiated by molecular methods or by crossing unifoliate materials with trifoliate orange (*Poncirus trifoliata*) pollen.

2. Morphology of seeds, embryos, and polyembryony

The study of polyembryony is receiving increased attention from both industrial and scientific sectors as it provides a form of cloning through seed that avoids the typical complications of sexual reproduction (for example, incompatibility barriers) and vegetative propagation (replication of viruses and other diseases) [20]. Another possible benefit of asexual reproduction by seed implies fixing the hybrid vigor and allowing the propagation of hybrids through many generations of seeds [21]. However, even in 2021, in [22] indicated that fixed hybrid vigor has not been possible because hybrid seeds cannot produce offspring with the same qualities.

Over the years, attempts to link the morphological characteristics of fruits, seeds, and embryos with polyembryony or with the characteristics of nucellar or zygotic embryos to predict the sexual or asexual origin of seedlings have taken place. Some researchers have studied the relationship between fruit weight and polyembryony in Swingle citrumelo, Volkamerian lemon, Cleopatra mandarin, and Amblicarpa [11, 23]. Other studies have tried to select characteristics of polyembryonic seeds as a possible method to anticipate the origin, zygotic or nucellar, of seedlings. For example, the size and shape classification of seeds has been associated with the production of zygotic seedlings on rootstocks [24, 25]. Also, logistic regression models that predict sexual seedlings have been developed [13].

Another variable that has been considered in the possible prediction of predominant sexual or asexual reproduction is the relationship between the percentage of polyembryony in citrus genotypes and the production of nucellar seedlings. Different studies have shown that species with a higher percentage of polyembryony are less likely to develop zygotic seedlings [21, 26, 27] because only one zygotic embryo possesses competitive disadvantage against all nucellar embryos: zygotic embryos tend to be small and do not survive in field conditions. In contrast, more numerous apomictic embryos tend to be large and produce more vigorous seedlings. One case is *Citrus reshni*: apomictic reproduction occurs based on the degree of polyembryony of the seeds (70–90%) and the small size (≤3 mm) of the sexual embryos [28]. However, this characteristic varies between genotypes; for example, in [12] reported six genotypes with a high percentage of polyembryonic seeds (65–98%), where only Valencia orange and Amblicarpa mandarin have the possibility of apomictic reproduction. Therefore, the presence and degree of polyembryony in the seed is a genetic characteristic dependent on the interaction with the environment, particularly with the weather, the stage of development of the plant and its organs, or the physiological conditions [29, 30]. Therefore, using the phenotypic characteristics of seeds, embryos, and polyembryony to identify the seedlings' sexual or clonal origin often leads to the omission of zygotic seedlings in segregating populations, which is unfavorable in breeding programs. This trait will be discussed later in the topic of molecular markers.

Other features of the embryos, such as size and location, have been studied to define the position and origin of nucellar and zygotic embryos. Zygotes often appear as more diminutive in seed size, with slower growth than adventitious embryos and growth prevailing at the micropylar end [31, 32]. However, during nucellar embryogenesis associated with the fertilized embryo sacs, there seems to be an inhibitory effect in the chalaza region [10, 33]. Consequently, both embryos preferentially grow towards the micropyle. These arguments are helpful to generate reproduction models in polyembryonic citrus; however, it should be noted that these studies focused on the initial stages of embryo and seed formation. Therefore, it is essential to develop innovative methods of identifying nucellar and zygotic embryos in mature seeds and differentiating the seedlings that originate in the early stages of development. We will cite the work of Andrade-Rodríguez et al. [28], who, using RAPD (Random

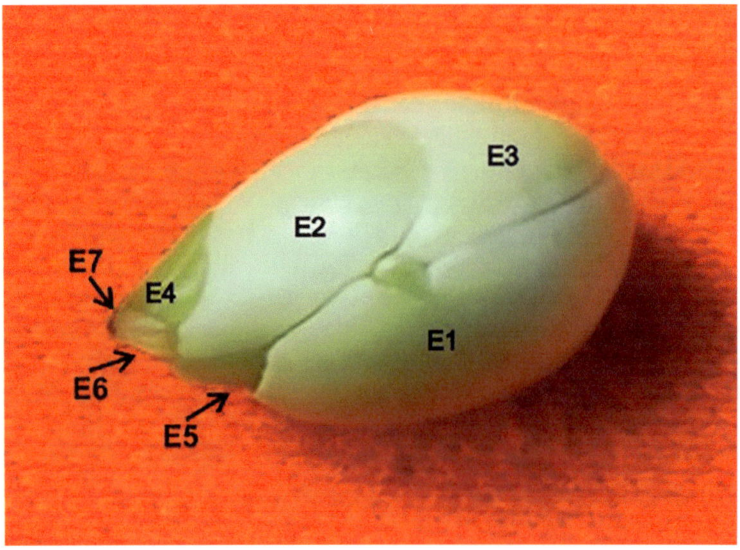

Figure 1.
Embryos classified by size in a seed of Amblicarpa mandarin (Citrus amblycarpa (Hassk) Ochse). E1 corresponds to the largest embryo in length and weight. E7 corresponds to the smallest. For each embryo, its location towards the micropyle or the chalaza and radicle orientation were recorded.

Amplification of Polymorphic DNA) markers, in vitro culture, and embryo morphology, concluded that not all small embryos located in the micropyle produce zygotic seedlings in Volkamerian lemon seeds. This contribution generated various questions about the reproduction models proposed for citrus and motivated us to pursue our current research focus.

Our laboratory has studied polyembryony in citrus from 2000 to date (as well as in other fruit trees not mentioned in this chapter). We have used molecular markers, in vitro culture of embryos separated by size, and grafting of the plants obtained in the greenhouse. These techniques have produced plants from embryos one, two, three, four, five, six, and seven (Example: E1, E2, E3, E4, E5, E6 and E7 in **Figure 1**) and propose new approaches to study the relationship between size and sexual, asexual, or different origin of the plants obtained from each of the embryos. Among the works published in citrus are references [7, 11, 12, 28, 34].

3. Identification of nucellar embryos and possible zygotes

Successfully harnessing apomixis for citrus propagation requires understanding its facultative nature; the formation of nucellar and zygotic embryos varies among environments and times of the year. Additionally, each species genetic makeup acts as an activation "switch" that relies on environmental changes. All these variables lead to variations in the degree of polyembryony, number of embryos per seed, and even the type of reproduction that will prevail, asexual or sexual [3, 30, 35]. Understanding the apomictic phenomenon in citrus is central to various studies to benefit from its dual characteristics: clonal propagation from seed and hybrids production. We seek to generate models that relate the characteristics of polyembryony, size, and location of the embryos in the seed, with their sexual or asexual origin. However, the selection of the hybrids that result from the cross with polyembryonic parents is complex if they do not express a reproducible dominant trait under several environmental conditions. In addition, the percentage of zygotic progenies in various citrus hybrids has been found to vary based on the seed parent used [36], the pollen origin [37], and environmental factors [38]. For this reason, researchers used various morphological, biochemical, and molecular markers to identify seedlings in the early stages of development.

Plant morphology is the first visible marker for hybrid selection in plant breeding. The first morphological trait used to differentiate nucellar plants from zygotic ones relied on linking vigor with asexual origin. For example, in 1949 in Ref. [39] published that nucellar plants are those that develop juvenile characteristics, such as vigorous growth, presence of thorns, or slow fruiting. Another morphological marker, which is also a dominant phenotypic trait, is leaf morphology [26]. In **Figure 2**, Amblicarpa nucellar plant (**Figure 2a**) and the hybrid that exhibits more than one leaflet (**Figure 2b**) are compared; both plants have thorns, which according to Cameron and Johnston [39] are juvenile characteristics. It has been used in nurseries since the first decades of the 20th century to identify hybrids by crossing trifoliate genotypes. The method is still valid and is used to obtain trifoliate rootstocks such as citrandarins (Cleopatra mandarin× P. trifoliata). However, both the presence of juvenile traits and multifoliate leaves are markers that vary depending on environmental conditions and plant development, so they are unreliable [40].

Not all morphological markers can reliably distinguish between zygotic and nucellar seedlings [40]. The expression of "trifoliate leaf" is a dominant trait over

Figure 2.
Mandarin Amblicarpa seedling (Citrus amblycarpa (Hassk) Ochse), obtained from embryos in vitro cultured, grafted on Volkamerian lemon (50 days after grafting). 2a nucellar plant and 2b sexual plant, both plants show "vigor" in height, leaf development, and thorn size; however, the sexual plant develops leaflets.

the recessive unifoliate leaf trait makes it easy to identify the first-generation hybrid seedlings in crosses between unifoliate citrus and trifoliate orange male parents [41]. Thus, hybrid seedlings with multifoliate leaves would be expected to appear when crosses of trifoliate male parents (*P. trifoliata*) are made. However, when Poncirus hybrids (such as citranges or citrumelos) are backcrossed to Citrus, seedlings with bi- or trifoliate leaves may vary considerably [37]. Furthermore, it is particularly difficult to remove off-type seedlings based on seedling morphology when pummelo has been used as a seed parent because the morphological characteristics associated with pummelo dominate the seedling phenotype [29]. An alternative to morphological markers is molecular markers, which can unequivocally discriminate the zygotic seedlings [28–30, 42, 43].

Biochemical tests have also been used as genetic markers in discriminating zygotic seedlings. Such is the case of colorimetry [44], chromatography [45], polyphenol darkening [46], spectroscopy [47], and isozyme analysis techniques [48]; however, these techniques fail to identify all clones and ignore some hybrids. For example, although they are codominant markers with simple methods, isoenzymes express few polymorphic loci to differentiate F1 seedlings from the female parent. Additionally, their expression level is qualitatively and quantitatively affected by environmental factors, plant development stage, or physiological conditions [29, 30].

Molecular biology expanded the tools available for identifying seedling sexual or asexual origin through molecular markers based on PCR (polymerase chain reaction). Among the most used markers are RAPD (Random Amplification of Polymorphic DNA), ISSR (Inter Simple Sequence Repeat), SSR (Simple Sequence

Repeat), and SNP (single-nucleotide polymorphism) [30, 34, 42, 43]. Various studies compare the efficiency of genetic markers in selecting hybrids, in Ref. [29] evaluated RAPD and EST-SSR (Expressed Sequence Tag-SSR), in [43] worked with morphological markers and SSR, and in [45] compared morphological markers and SNPs. RAPDs (dominant marker) and EST-SSRs (codominant marker) efficiently and accurately identified nucellar plants of mandarin (*C. reticulata* Blanco) and pumelo (*C. maxima* Meer.); both markers were highly related ($p = 0.001$) [29]. The expression of trifoliate leaves with molecular markers was complemented in Ref.s [43, 49]; their research showed the viability of early selection of hybrid seedlings based on the morphological marker and subsequent analysis with molecular markers in the seedlings that did not express multifoliate leaves. This procedure reduces the cost and time of the analysis.

Not only can genetic markers complement each other, but in vitro germination can also be used in embryo identification to increase the development of small embryos (which do not germinate under in vivo conditions). Similarly, faster acclimatization and growth of plants resulting from these embryos result from grafting on a rootstock. Three studies show the advantages of combining in vitro culture and identifying hybrids with SSR markers. It is worth mentioning that SSRs have been widely used to discriminate embryos according to their origin because they favor the selection of plants obtained by self-pollination and cross-pollination. Embryos from F1 seeds from a cross between 'Shiranuhi' mandarin and 'Shiranuhi' orange were cultivated *in vitro* to determine the percentage of zygotic embryos detected with SSRs depending on the days after pollination (90, 105, 125, 145 and 180 days, DAP) [50]. Growth in an artificial medium allowed them to maintain constant humidity and nutrient supplementation and achieved 36–75% germination depending on DAP, allowing identification of zygote embryos: 12% at 90 days, 8% at 105 days, 7% at 125 days, 1% at 145 days and 4% at 180 days after pollination. On the other hand, *in vitro* culture, SSR markers, and morphological markers have been used [49]; they first selected using a trifoliate leaf morphological marker and then, with SSRs, selected 41% of hybrid seedlings of rough lemon and citrandarin X-639c, and 46% of rough lemon and Swingle citrumelo hybrids. Likewise, these authors state that not all zygotic embryos survived until fruit maturation as the ratio of hybrid seedlings decreases with advancing stages of embryonic development after 95 days after pollination. Finally, *in vitro* culture, seedling grafting, and SSR markers have been used to identify nucellar embryos in citrange C-35, Amblicarpa mandarin, and Volkamerian lemon rootstocks, and Valencia orange and Minneola tangelo cultivars [12]. Also, this chapter includes the identification of nucellar embryos in Parson Brown orange, not published in the 2021 paper. This study differs in its approach as the authors only studied the largest embryo per seed, finding that the weight of the seed correlates with the weight of the largest embryo (0.76–0.96, $p \geq 0.05$). They used in vitro germination until they achieved 3–4 mm long seedlings and grafted them on Volkamerian lemon to promote their growth for 5 months until foliar collection. This procedure allowed for enough leaf material for DNA extraction (1.8–2.0 at 260/280 nm absorbance) without damage to the grafted plant, which developed under simulated nursery management. The SSR markers identified 70% nucellar seedlings in C-35, 65% in Volkamerian and Minneola, 85% in Amblicarpa and Valencia, and 100% in Parson Brown (**Figure 3**). Focus on the embryos with the most significant capacity to germinate in each seed allowed linking size (8–10 mm in length, 6–19 mg in weight) and the probability of generating nucellar plants; only Amblicarpa, Valencia, and Parson Brown showed a tendency to clonal propagation.

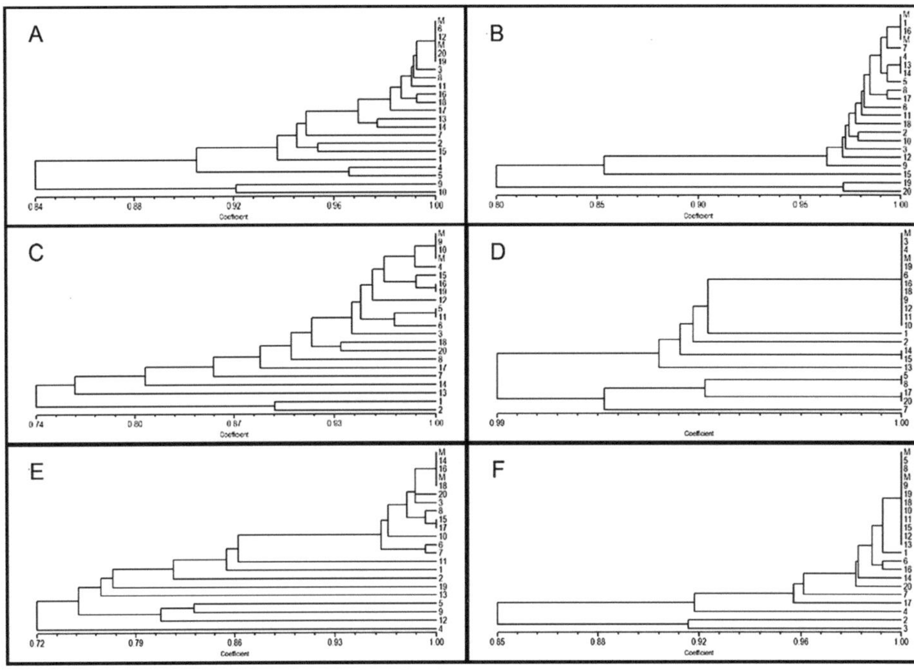

Figure 3.
Dendrograms showing the Genetic Similarity Indices (GSI) between nucellar seedlings and the female parent, obtained from the largest embryo per seed in six polyembryonic citrus cultivars (A, Citrange C-35; B, Amblicarpa Mandarin; C, Volkamerian lemon; D, Parson Brown orange; E, Tangelo Minneola; F, Valencia orange). The GSI was calculated with 30 SSR markers.

Various works that use molecular markers to classify seedlings according to their sexual or asexual origin state that those with total similarity to the female parent are nucellar. In [29] consider seedlings of Swingle citrumelo (*Citrus paradisi* Macf. × *Poncirus trifoliata* (L.) Raf.) and sour orange (*C. aurantium* L.) as "true nucellar seedlings" when they were identical to the female parent at all loci (100%), using EST-SSR and Euclidean distance with UPGMA cluster analysis. However, in [12] seedlings of six citrus genotypes with a Genetic Similarity Index [51] of 0.95 (95%) were considered "nucellar" compared to the female parent (**Figure 3**). Both studies show the discrepancy between the genetic similarity values needed for a plant to be considered "true nucellar" or "possible zygotic" to the female parent. As any marker has the disadvantage of characterizing only a tiny part of the genome, a discussion on the number of molecular markers needed for these studies is required. As an example, in [29] evaluated 12 EST-SSR markers, of which eight primer pairs revealed polymorphism, whereas in [12] examined 30 SSR markers, but only 17 were polymorphic enough to differentiate hybrids (TAA 41 and F4 being the most informative).

One of the objectives of breeding programs is to select plants based on agronomic and economic traits, such as the expression of polyembryony. However, it is relevant to consider the advantage of using molecular markers. These have the possibility of generating QTL (Quantitative Trait Loci) by associating the polymorphism to a phenotypic trait. The QTL Apo1, Apo2, Apo3, Apo4, and Apo6 associated molecular markers (RAPD, SSR, and CAP -Cleaved Amplified Polymorphic Sequence) with apomixis in the progeny of *C. volkameriana* × P. *trifoliata* cv 'Rubidoux.' *Apo2* is also associated with the activation of the type of embryo (mono-embryonic or poly-embryonic) in apomictic genotypes [52]. Further information has been obtained from AFLPs (Amplified Fragment Length Polymorphism) as marker loci have identified a genomic region associated with the percentage of polyembryonic seeds in *Citrus* × Poncirus hybrids [53]. The genetic control of adventitious embryology was also documented in *C. reticulata* [2, 22]; it has been proposed that the CiRKD1 gene regulates the somatic embryogenesis that occurs with two alleles that originate polyembryony and embryogenesis [54, 55]. So, the genetic control of apomictic reproduction and polyembryony is "more than a single switch" to activate it; it involves various genome regions in citrus.

Despite the effectiveness of molecular markers to discriminate between hybrids and nucellar plants, ongoing efforts try to relate the expression of morphological features of the embryo and its location in the seed with sexual origin. The purpose is to separate hybrid plants from clonal plants from seed or in the early stages of development without a laboratory. Various studies have related polyembryonic traits and origin; however, we will only analyze the *Citrus volkameriana* Pasq rootstock. The sexual origin was related to the embryo's size employing the trifoliate leaf marker expressed by the cross between the Volkamerian lemon and *P. trifoliata* [26]. They report large embryos (>5 mm) generate up to 82.7% of hybrid plants, while small embryos (1–2.9 mm) produce only 5.8%. Meanwhile, 25.9% of zygotic plants were identified with RAPDs in mature polyembryonic seeds, of which small embryos produced less than 43% (2–3 mm) and near the micropyle [28]. On the other hand, in Volkameriano seeds, when the largest embryo was separated from each polyembryonic seed and studied its location; all older embryos are in the chalaza and 45% of it produced non-clonal plants [12]. It is relevant to mention that the relationships between embryo morphology and sexual origin vary among genotypes [12, 26], strengthening the idea that embryos with a greater capacity to germinate in vivo in a polyembryonic genotype do not imply clonal reproduction. Thus, the competition caused by the asynchronous development between embryos during seed formation [7, 56] does not always favor the growth of nucellar embryos.

Therefore, molecular markers and complementary techniques (in vitro culture and grafting) are adequate to identify sexual and asexual seedlings with greater certainty than the exclusive use of morphological markers. Nevertheless, the characteristics of these techniques need to be assessed based on the objective and possibilities of the investigation. For example, although RAPDs and ESS-SSRs show a similar capacity to differentiate zygotic and nucellar seedlings [29], RAPDs are simpler and cheaper than EST-SSRs. However, the latter shows greater reproducibility among laboratories and detects all alleles of a locus (codominance). In the case of SNPs, their limitation is the cost and availability of the necessary equipment in the laboratory. Instead, in vitro culture is an excellent technique to be applied in research where germination of embryos is required for subsequent identification, but it is expensive in commercial citrus propagation schemes.

4. Conclusions

Polyembryony in citrus is complex and affected by genetic, physiological, and environmental factors. Therefore, a better understanding of the development of the embryos in the seed requires further studies. We propose combining complementary techniques to molecular markers to identify the sexual or asexual origin, the individual follow-up of each embryo in the seed, and the subsequent evaluation of morphological and production characteristics of the plant. This process in citrus requires approximately 3 and 4 years to start fruiting. However, it is necessary to answer a few questions: If a seedling identified with molecular markers as "different or possibly zygotic" (with similarity indices less than 0.95) expresses these polymorphisms phenotypically? Do the embryos, based on their size and position in mature seeds, tend to be nucellar or possibly zygotic? Is it convenient to continue considering the percentage of polyembryony in citrus as a form of clonal propagation in nurseries? Are the plants obtained from the larger embryo asexual? These and so many more questions need further research.

Acknowledgements

The authors thank Postgraduate College for the facillities made available during these studies. We also thank Vivero Cazones for providing the plant material and plant care help. This work was supported by the Mexican Council of Science and Technology (CONACyT).

The authorship criteria are listed in our Authorship Policy: https://www.intechopen.com/page/authorship-policy.

This section of your manuscript may also include funding information.

Conflict of interest

The authors declare no conflict of interest.

Author details

Angel Villegas-Monter[1]*, Elisa Del Carmen Matínez-Ochoa[2],
María Andrade-Rodriguez[3] and Itzel Villegas-Velázquez[4]

1 Department of Plant Physiology, Colegio de Postgraduados,
Texcoco, Estado de México, México

2 Technological Highschool Center No. 3, Texcoco, Estado de México, México

3 Faculty of Agricultural Science, Universidad Autónoma de Morelos,
Cuernavaca, Morelos, México

4 Department of Botany, Colegio de Postgraduados, Texcoco, Estado de México,
México

*Address all correspondence to: villema53@hotmail.com

References

[1] Albertini E, Barcaccia G, Carman JG, Pupilli F. Did apomixis evolve from ser or was it the other way around? Journal of Experimental Botany. 2019;**70**:2951-2964. DOI: 10.1093/jxb/erz109 DOI: 10.1093/jxb/erz109

[2] Schmidt A. Controlling apomixis: Shared features and distinct characteristics of gene regulation. Genes. 2020;**11**:329. DOI: 10.3390/genes11030329

[3] Hand ML, Koltunow AM. The genetic control of apomixes: Asexual seed formation. Genetics. 2014;**197**(2):441-450

[4] Whitton J, Sears CJ, Baack EJ, Ottoy SP. The dynamic nature of apomixis in the angiosperms. International Journal of Plant Science. 2008;**169**:169-182. DOI: 10.1086/523369

[5] Carneiro VTC, Dusi DMA., Ortiz JPA. Apomixis: Ocurrence, Aplications and Improvements. In: Teixeira da SJA, editors. Floriculture, Ornamental and Plant Biotechnology. Vol 1. Global Science Books. 2006. p. 564-571

[6] Bicknell RA, Koltunow AM. Understanding apomixis: Recent advances and remaining conundrums. The Plant Cell. 2004;**16**(1):S228-S245. DOI: 10.1105/tpc.017921

[7] Sánchez-Damas JJ, Avitia EG, Castillo GAM, Villegas AM, Corona TT. Estudio anatómico de la poliembrionía en tres portainjertos de cítricos. Revista Chapingo. Serie horticultura. 2006;**12**(2):145-152

[8] Kishore K, Monika N, Rinchen D, Lepcha B, Pandey B. Polyembryony and seedling emergence traits in apomictic citrus. Scientia Horticulturae. 2012;**138**:101-107. DOI: 10.1016/j.scienta.2012.01.035

[9] Naumova TN. Apomixis in Angiosperms: Nucellar and Integumentary Embryony. Mershchikova I., Trans. 1st ed. Boca Raton, Florida, USA: CRC Press; 1993. DOI: 10.1201/9781351069755

[10] Koltunow AM. Apomixis: Embryo sacs and embryos formed without meiosis or fertilization in ovules. The Plant Cell. 1993;**5**:1425-1437. DOI: 10.1105/tpc.5.10.1425

[11] Andrade-Rodríguez M, Villegas-Monter A, García-Velázquez A. Características morfológicas del fruto y poliembrionía de tres portainjertos de cítricos. Revista Chapingo Serie Horticultura. 2003;**9**(2):255-263

[12] Martínez-Ochoa E d C, Villegas-Velázquez I, Alarcón-ZúñigaB,González-HernándezVA, Villegas-Monter A. Polyembryony in citrus: Does the largest embryo in the seed develop a nucellar seedling? Scientia Agricola. 2021;**79**(6):e20200060. DOI: 10.1590/1678-992x-2020-0060

[13] Bowman KD, Gmitter FG, Hu X. Relationships of seed size and shape with polyembryony and the zygotic or nucellar origin of Citrus spp seedlings. HortScience. 1995;**30**(6):1270-1282. DOI: 10.21273/HORTSCI.30.6.1279

[14] Sampaio PO, Santos LP, Costa dos Santos L, Correa RC, Santos WSF. Characterization of Poncirus trifoliata hybrids and other citrus rootstocks in the State of Bahia (Caracterização de híbridos de Poncirus trifoliata e de outros porta-enxertos de citros no Estado da Bahia). Revista Brasileira de Fruticultura.

2006;**28**(3):410-413. DOI: DOI10.1590/So100-29452006000300016

[15] Da Silva Rodrigues MJ, Da Silva Ledo CA, Girardi EA, Da Hora Almeida LA, Filho WSS. Fruit characterization and propagation of hybrid citrus rootstocks in protected environment. [Caracterização De Frutos E Propagação De Porta-Enxertos Híbridos De Citros Em Ambiente Protegido]. Revista Brasileira De Fruticultura. 2015;**37**(2):457-470. DOI: 10.1590/0100-2945-068/14

[16] Aleza P, Juárez J, Ollitrault P, Navarro L. Polyembryony in non-apomictic citrus genotypes. Annals of Botany. 2010;**106**(4):533-545. DOI: /10.1093/aob/mcq148

[17] Zhang S, Liang M, Wang N, Xu Q, Deng X, Chai L. Reproduction in woody perennial Citrus: An update on nucellar embryony and self-incompatibility. Plant Reproduction. 2018;**31**:43-57. DOI: 10.1007/s00497-018-0327-4

[18] Cook MT. Notes on polyembryony. Torreya. 1907;**7**(6):113-117 Available from: http://www.jstor.org/stable/40594604

[19] Strasburger EU. Polyembryonie. Jenaisch. Zeitsch. Naturwiss. 1878;**12**:647-670

[20] Barcaccia G, Albertini E. Apomixis in plant reproduction: A novel perspective on an old dilemma. Plant Reproduction. 2013;**26**(3):159-179. DOI: 10.1007/s00497-013-0222-y

[21] Koltunow AM, Scott NS, Chaudhury AM. The use of apomixis in cloning horticultural plants. Current applications and molecular prospects. Acta Horticulturae. 2001;**560**:333-343. DOI: 10.17660/ActaHortic.2001.560.64

[22] Fiaz S, Xiukang W, Younas A, Alharthi B, Riaz A, Ali H. Apomixis and strategies to induce apomixis to preserve hybrid vigor for multiple generations. GM Crops and Food. 2021;**12**:57-70. DOI: 10.1080/21645698.2020.1808423

[23] Darlan-Ramos J, Neto SEA, Castro NEA, Martis PCC, Correia MG. Poliembrionia e caracterizaçao de frutos de citrumelo 'Swingle' e de Poncirus trifoliata. Ciência e Agrotecnologia. 2006;**30**(1):88-91. DOI: 10.1590/S1413-70542006000100012

[24] Duarte FEV d O, Barros DDR, Girari EA, Soares-Filho WDA, Passos OS. Poliembrionia e atributos morfológicos de sementes de porta-enxertos de citros. Revista Brasileira de Fruticultura. 2013;**35**(1):246-254. DOI: 10.1590/S0100-29452013000100028

[25] Xiang C, Roose ML. Frequency and characteristics of nucellar and zygotic seedling in 12 citrus rootstocks. Scientia Horticulturae. 1988;**37**(1-2):47-59. DOI: 10.1016/0304-4238(88)90150-1

[26] Soares-Filho WDS, Moreira CS, Da Cunha MAP, Da Cunha-Sobrinho AP, Passos OS. Poliembrionia e freqüência de híbridos em Citrus spp. Pesquisa Agropecuaria Brasileira;**200**(35):857-864. DOI: 10.1590/S0100-204X2000000400025

[27] Soost RK, Roose ML. Citrus, chapter 6. In: Janick J, Moore JN, editors. Fruit Breeding Vol I: Tree and Tropical Fruits. Wiley: New York; 1996. pp. 257-323

[28] Andrade-Rodríguez M, Villegas A, Carrillo G, García A. Polyembryony and identification of Volkamerian lemon zygotic and nucellar seedlings using RAPD. Pesquisa Agropecuária Brasileira. 2004;**39**(6):551-559. DOI: 10.1590/S0100-204X2004000600006

[29] Rao MN, Soneji JR, Chen C, Huang S, Gmitter FG. Characterization of zygotic

and nucellar seedlings from sour orange-like Citrus rootstock candidates using RAPD and EST-SSR markers. Tree Genetics & Genomes. 2008;**4**:113-124. DOI: 10.1007/s11295-007-0092-2

[30] Yildiz E, Kaplankiran M, Demirkeser TH, Uzun A, Toplu C. Identification of zygotic and nucellar individuals produced from several citrus crosses using SSRs markers. Notulae Botanicae Horti Agrobotanici. 2013;**41**(2):478-484

[31] Frost HB, Soost RK. Seed Reproduction: Development of Gametes and embryos, In: the citrus industry, Volumen 2. Reuther, W., I. D. Batchelor, H. J. Webber (eds.). Division of Agriculture and Science, University of California Press. Berkeley, Cal., USA; 1968. pp. 290-324

[32] Thakur DR, Bojwa BS. Extent of polyembryony in some species and varieties of Citrus. Indian Journal Horticulturae. 1971;**28**(1):25-28

[33] Wakana A, Uemoto S. Adventive embryogenesis in Citrus (Rutaceae). II Postfertilization Development. American Journal of Botany. 1988;**75**:1033-1047. DOI: 10.1002/j.1537-2197.1988.tb08810.x

[34] Andrade-Rodríguez M, Villegas-Monter A, Gutiérrez MA, Carrillo G, García A. Polyembryony and RAPD markers for identification of zygotic and nucellar seedlings in Citrus. Agrociencia. 2005;**39**:371-383

[35] Grossniklaus U, Nogler GA, Van Dijk PJ. How to avoid sex: The genetic control of gametophytic apomixis. The Plant Cell. 2001;**13**:149-1497

[36] Spiegel-Roy P, Bardi A, Shani A. Peroxidase isozymes as a tool for early separation of nucellar and zygotic citrus seedlings. Second Meeting of the International Society of Citriculture. Proceeding of the International Society of Citriculture. Vol. 2. Florida, USA. 1977. pp. 619-624

[37] Cameron JW, Soost RK. Leaf types of F1 hybrids and backcrosses involving unifoliate Citrus and trifoliate Poncirus. Journal of the American Society for Horticultural Science. 1980;**105**:517-519

[38] Moore GA, Castle WS. Morphological and isozymic analysis of open-pollinated Citrus rootstock populations. Journal of Heredity. 1988;**79**:59-63. DOI: 10.1093/oxfordjournals.jhered.a110448

[39] Cameron JW, Jonhston JC. Nucellar Seedlings may permit development of disease-free citrus varieties. California Agriculture. 1949;**3**(1):9-12

[40] Caruso M, Distefano G, Paolo DP, Malfa SL, Russo G, Gentile A, et al. High resolution melting analysis for early identification of citrus hybrids: A reliable tool to overcome the limitations of morphological markers and assist rootstock breeding. Scientia Horticulturae. 2014;**180**:199-206. DOI: 10.1016/j.scienta.2014.10.024

[41] Sykes SR. Characterization of citrus rootstock germplasm introduced as seeds to Australia from the People's Republic of China. Scientia Horticulturae. 2011;**127**:298-304. DOI: 10.1016/j.scienta.2010.10.015

[42] Golein B, Fifaei R, Ghasemi M. Identification of zygotic and nucellar seedlings in citrus interspecific crosses by inter simple sequence repeats (ISSR) markers. African Journal Biotechnology. 2011;**10**:18965-18970

[43] Zhu S, Wu B, Ma Y, Chen J, Zhong G. Obtaining citrus hybrids by in vitro culture of embryos from mature seeds and early identification of hybrid

seedlings by allele-specific PCR. Scientia Horticulturae. 2013;**161**:300-305. DOI: 10.1016/j.scienta.2013.07.003

[44] Furr JR, Reece PC, Hornciar G. Identification of hybrid and nucellar citrus seedlings by a modification of the rootstock color test. In Proceedings of American Society for Horticultural Science. 1946;**48**:141-146

[45] Tatum JH, Berry RE, Hearn CJ. Characterization of Citrus cultivars and separation of nucellar and zygotic seedlings by thin layer chromatography. Proceedings Florida State Horticulture Society. 1974;**87**:75-81

[46] Esen A, Soost RK. Inheritance of browning of young-shoot extracts of citrus. The Journal of Heredity. 1974;**65**:97-100

[47] Pieringer AP, Edwards GJ. Identification of nucellar and zygotic citrus seedlings by infrared spectroscopy. Journal of American Society for Horticultural Science. 1965;**86**:226-234

[48] Anderson CM, Castle WS, Moore GA. Isozymic identifcation of zygotic seedlings in Swingle citrumelo Citrus paradise × Poncirus trifoliata nursery and field populations. Journal American Society for Horticultural Science. 1991;**116**:322-326. DOI: 10.21273/JASHS.116.2.322

[49] Singh J, Dhaliwal HS, Thakur A, Sidhu GS, Chhuneja P, Gmitter FG Jr. Optimizing recovery of hybrid embryos from interspecific citrus crosses of polyembryonic rough lemon (Citrus jambhiri Lush.). Agronomy. 2020;**10**:1940. DOI: 10.3390/agronomy10121940

[50] Kim M, Kim SH, Kim HB, Park YC, Song KJ. 2020. Some factors affecting the efficiency of hybrid embryo rescue in the 'Shiranuhi' Mandarin. Horticultural Science and Technology. 2020;**38**(2):271-281. DOI: 10.7235/HORT.20200026

[51] Mannen H, Tsuji S, Mukai F, Goto N, Ohtagaki S. Genetic similarity using DNA fingerprinting in cattle to determine relationship coefficient. Journal of Heredity. 1993;**84**(3):166-169. DOI: 0.1093/oxfordjournals.jhered.a111310

[52] García MR, Asins MJ, Forner J, Carbonell EA. Genetic analysis of apomixis in Citrus and Poncirus by molecular markers. Theoretical and Applied Genetics;**199**(99, 511):-518. DOI: 10.1007/s001220051264

[53] Kepiro JL, Roose ML. AFLP markers closely linked to a major gene essential for nucellar embryony (apomixis) in Citrus maxima × Poncirus trifoliata. Tree Genetics & Genomes. 2010;**6**:1-11. DOI: 10.1007/s11295-009-0223-z

[54] Wang X, Xu Y, Zhang L, Cao Y, Huang Cheng J. Genomic analyses of primitive, wild and cultivated citrus provide insights into asexual reproduction. Nature Genetics. 2017;**49**:765-772. DOI: 10.1038/ng.3839

[55] Shimada T, Endo T, Fujii H, Nakano M, Sugiyama A, Daido G, et al. MITE insertion-dependent expression of CitRKD1 with a RWP-RK domain regulates somatic embryogenesis in citrus nucellar tissues. BMC Plant Biology. 2018;**18**:166. DOI: 10.1186/s12870-018- 1369-3

[56] Koltunow AM, Soltys K, Nito N, McClure S. Anther, ovule, seed, and nucellar 343 embryo development in Citrus sinensis cv. Valencia. Canadian Journal Botany. 1995;**73**(344):1567-1582. DOI: 10.1139/b95-170

Chapter 4

Quality of Postharvest Degreened Citrus Fruit

Julia Morales, Lourdes Cervera, Pilar Navarro and Alejandra Salvador

Abstract

External color is a key factor that defines external citrus fruit quality. Degreening with exogenous ethylene exposure is a widely used postharvest treatment applied to promote external citrus fruit color development, mainly with those cultivars that reach internal maturity while their external peel color is still green. Ethylene plays a crucial role in the color change of citrus fruit because it induces two simultaneous, but independent, processes—chlorophyll degradation and carotenoid synthesis. However, it is important to know, in addition to the effect on skin color development, whether this treatment can negatively affect other fruit quality parameters. This chapter addresses the influence of postharvest degreening treatment on the physico-chemical, nutritional, and sensory quality of citrus fruit.

Keywords: ethylene, color, biocomponents, antioxidant, disorders, flavor

1. Introduction

External citrus fruit color is an important quality attribute to be considered for the fresh market. Consumers associate high-quality fruit with the typical bright orange skin color, while they associate greenish skins with unripe fruit [1].

In citrus fruit, color change requires not only night temperatures close to 12°C but also marked differences between day and night temperatures. At 12°C, the expression of the key genes of the carotenoid biosynthesis pathway is stimulated, which increases concentrations of carotenoids and xanthophylls [2]. However, in tropical countries without cold night temperatures, citrus fruit reaches acceptable internal quality without displaying the characteristic orange color, and fruit is consumed when the skin is yellowish-green.

In Mediterranean citrus production areas, early citrus varieties, especially mandarins and oranges, reach internal maturity before full external coloration. In these cases, a degreening treatment, with the application of exogenous ethylene, is a common postharvest practice followed to improve external fruit color, which contributes to their market acceptance and extends their marketing season. Moreover, in mature citrus fruit harvested after the color break, it has been reported that exogenous ethylene can reduce the incidence of important postharvest physiological disorders, such as non-chilling peel pitting or chilling injury [3, 4]. These disorders are manifested as peel damage, as well as the incidence of the disease caused by *Penicillium digitatum* without negatively affecting fruit organoleptic properties [3, 5].

The degreening treatment applied to enhance external citrus fruit color involves exposing the fruit to low ethylene concentrations [6]. However, in addition to ethylene concentration, other important factors, such as temperature, humidity or carbon dioxide (CO_2), and oxygen in the atmosphere, are implied during color change. The exhaustive control of all these factors, as well as the process duration, are requirements to achieve the desired external color without promoting undesirable reactions related to physiological disorders or sensory changes, which harm fruit quality during their posterior shelf-life.

This chapter describes the parameters involved in the degreening process, and its impact on external or internal fruit quality.

2. Factors involved in the postharvest degreening process

In climacteric fruit, ethylene plays a key role in governing physiological and biochemical changes that occur during ripening, including the color break, softening, and accumulation of sugars, acids, aroma volatiles, and vitamins [7]. In contrast, citrus fruit is non-climacteric, and their natural ripening is not accompanied by rises in respiration and ethylene production rates [8]. However, exposure to exogenous ethylene has been shown to stimulate various ripening-related processes, such as destruction of the green chlorophyll pigments and accumulation of orange/yellow carotenoids, in citrus peel tissue. Exogenous ethylene increases chlorophyllase activity and gene expression as well as other genes involved in chlorophyll breakdown [9, 10]. Ethylene also downregulates chlorophyll biosynthesis, by repressing the gene expression of Mg-chelatase and most genes involved in photosynthesis and chloroplast biogenesis [11]. Ethylene also stimulates the transcription of carotenoid biosynthetic genes in citrus fruit peel, which is concomitant to both the transformation from chloroplast into chromoplast and the accumulation of xanthophylls and carotenoids [11, 12]. Therefore, degreening treatment with exogenous ethylene exposure is a widely used postharvest treatment to promote external color development in citrus, especially in those cultivars that reach internal maturity while external peel color is still green [12, 13].

Commercial postharvest degreening is usually carried out in packinghouses, specifically in temperature-controlled chambers equipped with automatic injectors to provide the appropriate ethylene concentration. The applied ethylene concentration is low, close to 1–5 ppm [14, 15]. This concentration suffices to cause a color change, and it has been reported that increasing ethylene concentration has no significant effect on improving peel color or reducing degreening times [16]. Exposing citrus fruit to higher ethylene concentrations can cause undesirable effects related to accelerated fruit senescence. It is, therefore, necessary to monitor ethylene levels constantly throughout the process to ensure that its concentration is sufficient for proper degreening without detriment to quality.

Temperature is a determinant of color evolution during the ethylene degreening process. Temperature strongly influences chlorophyll degradation and carotenoid synthesis. High temperatures close to 30°C lead to rapid chlorophyll loss but delay carotenoid accumulation. However, the temperature within the 18–20°C range allows greater carotenoid accumulation, although chlorophyll degradation is slower. Very low temperatures (close to 5°C) during the process can repress carotenoid accumulation and affect carotenoid composition in flavedo [17]. In Spain, mandarins are subjected to degreening treatment at 18–21°C, while oranges are exposed to a slightly higher temperature of 20–22°C [18]. However, the degreening of lemons is carried out at 25–30°C [19]. Similarly in Israel and California, mandarins and oranges are exposed to 20–25°C [20–22]. In Florida, most citrus fruits are commercially degreened at a relatively high temperature of 28–29°C [23].

Another important factor to consider in the degreening treatment is the time required to reach the desired fruit color, which very much depends on both the cultivar and the initial fruit color which, in turn, are controlled by fruit maturity and orchard conditions [18]. Citrus peel color increases with the ethylene exposure time during the degreening process. However, the negative effects induced by ethylene are also stronger the longer the exposure time is. Therefore, the time during which fruit is exposed to ethylene should be as short as possible, and optimal temperature, ethylene concentration, humidity, and aeration condition should apply. Not exceeding 72–96 h of treatment is advisable to avoid peel disorders during posterior commercialization [24]. Current color sorters, which work with photoelectric cells on handling lines, allow the fruit to be selected according to their initial color and to adjust degreening treatment duration to avoid overexposing fruit to ethylene. It should be noted that fruit color evolution continues once the fruit is transferred from ethylene chambers to marketable conditions [6]. However, the temperature to which fruit is subjected after the degreening treatment is key for posterior color evolution. Color development can be limited if temperatures are low while shipping citrus fruit [25]. The combination of periods with and without ethylene exposure has successfully achieved the desired change in mandarins and oranges, and optimal ethylene exposure duration has been estimated by considering color at harvest and subsequent marketing conditions [6]. In addition, the color change that fruit undergoes during degreening treatment very much depends on the variety [6, 24, 26].

Exposing fruit to ethylene increases the fruit respiration rate inside degreening chambers [27]. CO_2 is known as an ethylene antagonist which, at high concentrations, inhibits the action of ethylene and delays the color change process. Therefore, the chambers in which degreening is carried out must be equipped with specific CO_2 sensors to continuously monitor its concentration. CO_2 must be kept below 0.15–0.2% to allow proper degreening. High CO_2 levels in the atmosphere may induce acetaldehyde and ethanol production with the consequent risk of off-flavors in fruit [20]. The oxygen concentration has to remain above 20% because, apart from its role in respiration activity, oxygen is necessary for chlorophyll degradation and carotenoid biosynthesis. Therefore, adequate ventilation is essential to supply oxygen and remove CO_2 accumulation in degreening chambers. Moreover, during degreening treatments, keeping relative humidity at around 95% is desirable to obtain satisfactory peel color change results and to avoid fruit dehydration and skin alterations [28].

3. Physiological disorders

Apart from the benefits that exogenous ethylene application during degreening treatment can provide to improve fruit coloration, it can lead to physiological alterations in fruit if the process is not properly carried out.

The negative effects associated with exposing the fruit to exogeneous ethylene are related to an accelerated senescence process, which can lead to major quality loss and a shorter shelf-life [26, 29]. These undesirable effects depend mainly on the concentration and duration of fruit exposure to ethylene, but also affect atmospheric conditions, such as low humidity, excessive temperature, and high CO_2 concentrations. The incidence of these alterations also depends on the cultivar, preharvest conditions, and the treatment of the postharvest condition. In most cases, these disorders do not appear immediately after the degreening process but are manifested later during storage until marketing [16]. Special care must be taken when fruits are to be shipped at low temperatures, which occurs with exports to countries with quarantine requirements because low temperatures after degreening can accelerate the manifestation of these alterations. Therefore, to avoid possible fruit disorders, treatment must be carried out under careful conditions by considering all the factors involved in the process, as well as the conditions to which fruit are subjected after degreening.

The most frequent physiological disorders associated with incorrect degreening treatment are described below.

3.1 Calyx senescence

Calyx drop and browning are the main physiological disorders associated with incorrect degreening (**Figure 1**). Keeping the calyx fresh during the postharvest life of citrus fruit is a quality requirement because fruits are commercially more attractive, and it also protects the fruit from fungal infection upon abscission [30].

Ethylene is a vegetal hormone that promotes tissue abscission and senescence in fruit [31]. Low atmospheric ethylene levels (≤ 0.01 μL/L) reduce this disorder during long-term storage, and its accumulation during cold storage has been related to calyx senescence [32].

During degreening treatment, ethylene application has been reported to increase the activity of both polygalacturonic acid enzyme (PG) and cellulase (Cx), which promote browning and calyx abscission [33]. The incidence of calyx senescence triggered by degreening depends mainly on the ethylene concentration and process duration [26, 30, 34], and it is also very cultivar dependent. Clementine mandarins 'Marisol' and 'Oronules' have been shown to be more sensitive to calyx senescence than 'Clemenpons' and 'Clemenules' [35]. Other citrus cultivars, such as 'Satsuma' mandarin and 'Navelina' orange, are extremely sensitive to calyx senescence [6, 26]. Despite these calyx disorders appearing sometime after degreening, their incidence is accentuated with post-treatment storage time.

Figure 1.
Calyx senescence. Browning and drop of the calyx.

Many studies have focused on finding solutions to prevent calyx senescence during degreening. Using auxins has been one of the effective treatments in controlling this alteration. Synthetic auxin 2,4-dichlorophenoxyacetic acid (2,4-D) has been extensively reported as a treatment to retard calyx abscission, drying, and browning. It occurs as a consequence of exposing the fruit to ethylene during degreening treatment [30, 34, 36]. This auxin's mechanism of action is to reduce PG and Cx activities and to increase the lignin and water contents of fruit peel [37]. Despite this auxin having been widely employed and proving useful, it has been restricted by current European Union legislation, hence the need to find other synthetic auxins to avoid physiological changes in the calyx [33, 35].

Carvalho et al. [35] tested four synthetic auxins (2,4-dichlorophenoxy propionic acid (2,4-DP); 3,5,6-trichloro-2-pyridyloxyacetic acid (3,5,6-TPA); 2,4-D isopropyl ester; 2,4-D-amine) for retarding the calyx disorders caused by degreening in different clementine cultivars. All the evaluated auxins reduced calyx senescence. In all the cultivars, the best results were obtained by applying 3,5,6-TPA, followed by 2,4-D isopropyl ester. In contrast, the treatment with 2,4-DP had a positive effect on avoiding calyx disorders, but only in 'Clemenpons' mandarins, and this effect was not observed in the other studied cultivars. Although auxin treatments can delay color evolution, they had no negative effect from a commercial point of view in all the studied cultivars because, after degreening, all the treatments had a commercially acceptable color index. Moreover, no auxin treatment affected the sensory quality of degreened fruit.

The application of 3,5,6-TPA at different concentrations (10, 20, and 40 ppm) under commercial degreening conditions has been well-studied with many varieties. The higher the 3,5,6-TPA dose, the lower the percentage of affected fruit with calyx alteration symptoms [38, 39]. In fact, 3,5,6-TPA is currently used in the postharvest industry at a dose of 40 ppm and is applied with drenchers [39].

Other plant growth regulators, such as HF-Calibra® (SIPCAM INAGRA, Spain) with active ingredient MCPA-thioethyl (S-Ethyl-4-chloro-o-tolyloxythioacetate), have been tested in the most important Spanish early-season citrus varieties subjected to degreening treatment at different concentrations (10, 20, 40, and 60 ml/L) [26]. That study revealed that this auxin contributes to decreasing calyx senescence triggered by exogenous ethylene. The higher the applied dose, the stronger the effect on preventing calyx abscission.

Another technique to avoid the negative effect of ethylene on calyx senescence is degreening treatments which combine different exposure periods with and without ethylene. This has been evaluated with mandarins and oranges to be exported to the USA from Spain [6, 24]. Optimal degreening process conditions have been established according to both the initial external color and variety [40]. These recommendations are provided in **Table 1**.

Recently, treatments with Oligochitosan and Chitosan (poly-β-(1,4)-D-glucosamine) have been demonstrated to reduce calyx browning caused by degreening treatment, which has been linked with the inhibition of protopectin, cellulose, and lignin degradation [33].

3.2 Peel disorders

Some citrus cultivars with very thin skin, especially mandarins, can manifest bruising symptoms when they pass along packing lines after being degreened.

Initial Citrus Color Index (CCI = 1000*a/L*b)	EU		USA-Japan	
	Mandarins	Oranges	Mandarins	Oranges
CCI < −13	Not recommended			
−13 > CCI < −5	72 h with ethylene	Not recommended	48–72 h with ethylene	Not recommended
−5 > CCI < +3	48 h with ethylene and 72 h without ethylene	72 h with ethylene	48–72 h without ethylene	48–72 h with ethylene
CCI > +3	24 h with ethylene and 48 h without ethylene	48 h with ethylene and 72 h without ethylene	24 h without ethylene	24 h with ethylene and 48 h without ethylene
CCI > +7	Suitable color. Degreening treatment is not necessary			

Table 1.
Recommendations for the degreening treatment of Spanish mandarins and oranges to be exported to the EU, USA, or Japan (adapted from Pássaro et al. [40]).

This disorder is commonly called 'zebra skin' and is caused by mechanical abrasion during the brushing or rolling of fruit in an equatorial area and the cells darken and produce necrotic streaks in the center of fruit segments (**Figure 2**) [41]. This disorder can also be manifested in non-degreened fruit, but degreening treatment enhances susceptibility to bruising. When degreening treatment is carried out at low humidity, elevated temperature, and high CO_2 levels, and for long ethylene exposure times, the susceptibility of fruit showing zebra skin symptoms after packing increases [42]. To avoid this peel alteration, keeping fruit in an ethylene-free atmosphere for at least 12 h after degreening and before packaging, and delaying harvest for 5–7 days after rainy periods, are highly recommended because skin turgidity can increase this damage [43].

Oleocellosis is one of the main postharvest skin disorders to occur in citrus. It is usually caused by mechanical injuries to cells in rind at harvest or on packing lines. Broken cells release the oil, which is phytotoxic to pericarp cells and causes browned areas on affected rind areas. Early-season citrus fruit is more susceptible to oleocellosis [44].

When fruit is subjected to degreening treatment, the mechanical damage caused at harvest and while transporting fruit from fields to packinghouses can lead to olleocelosis. Damaged cells, in which oil extravasation occurs, do not change color during degreening treatment, which leaves green areas on the rind (**Figure 3**). Moreover, the more turgid rind cells are, the more susceptible fruit are to mechanical damage and they, consequently, show oleocellosis. High environmental humidity, rainy days, or excess irrigation increase fruit peel turgidity, and it is advisable to avoid harvesting after rain or picking fruit early in the morning to avoid dew. Waiting for 24 h from harvesting before handling in containers is also recommended [45]. In addition, cultivar can be a relevant factor since as higher is the density of oil cavities, greater volume, and their position in the pericarp higher is the incidence of the disorder [46].

Another physiological disorder that can be accentuated by postharvest degreening is stem-end rind breakdown (SERB).

Figure 2.
'Zebra skin' peel disorder caused by mechanical abrasion in packing line.

Figure 3.
Oleocellosis symptoms after degreening treatment.

SERB symptoms cause the peel tissue around the calyx to collapse, which becomes dark and sunken (**Figure 4**). SERB is due to excessive water loss and usually appears 2 and 7 storage days after packaging. The fruit that develops this disorder tends to rot more easily. Thinner-skinned fruit grown in humid growing environments, such as Florida, tend to be more prone to SERB than thicker-skinned fruit from arid environments [47]. Including a holding period for 12 to 24 h after degreening treatment and before packing is recommended to avoid SERB. Beneficial effects of ethylene on fruit have been reported; for example, ethylene has been related as a contributor to new wax formation. It also increases the total soft epicuticular wax

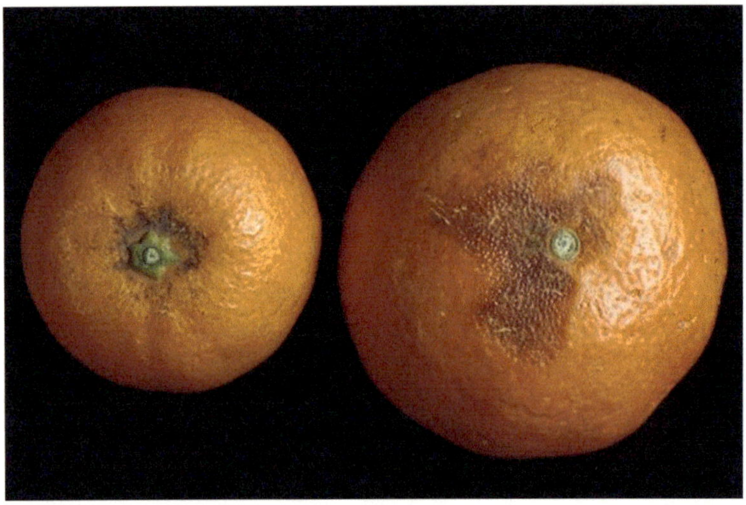

Figure 4.
Stem-end rind breakdown (SERB) damage.

content in mature citrus fruit, which decreases transpiration, maintains the water balance, and regulates the gas exchange in plants [48].

3.3 Pathological disorders

It has been reported that degreening can aggravate the incidence of some fungi. However, the effect of treatment depends on process conditions. The effect of degreening on *Penicillium* spp. development depends on the temperature at which treatment takes place. When degreening is performed at 20–22°C (Mediterranean Region conditions), the *Penicillium* incidence rises because these temperatures are close to optimal for this fungus to develop [21]. Nevertheless, when treatment is carried out at high temperatures (28–29°C) as in Florida, the action of this fungus is inhibited [49]. In this case, treatment can even be seen as curing, which lowers the *Penicillium* incidence. It has also been reported that *Penicillium* incidence in degreened fruit is maturation-dependent, and more mature fruit generally show greater severity [50]. After degreening, traditionally the selection of the fruit affected by *Penicillium* has been manually done, but artificial vision and hyperspectral imaging methods have been recently developed for the early detection of affected fruit [51]. In addition to sorting and discarding affected fruit, the application of washing, sanitation, and specific antifungal treatments before degreening is highly recommended.

Degreening may enhance the anthracnose incidence because ethylene induces conidia germination, appressoria formation, and *Colletotrichum gloeosporioides* germination due to more accessible nutrients on fruit surfaces [52]. The development of this disease depends on ethylene concentration and treatment duration [53].

Another disease associated with degreening is stem-end rot (SER) caused by *Diplodia natalensis*. SER develops from latent infections by this fungus established in necrotic fruit button tissue. After harvest, this fungus reactivates and infects fruit through the natural openings that form between fruit and buttons during button separation upon abscission [54]. The SER incidence can be enhanced by degreening treatment when the temperature comes close to 30°C and the ethylene concentration is high [37].

4. Effect of degreening on bioactive compounds

Nowadays, consumers demand fruit with, in addition to an attractive appearance, high sensory and nutritional quality. Citrus fruits are known for their health-promoting benefits due to their high content of bioactive compounds with antioxidant properties. The antioxidant activity of citrus fruit is due to the water-soluble fraction, including vitamin C and polyphenols, and also to the apolar fraction that contains carotenoids. Despite the common dogma that ethylene has a minor effect on the internal ripening processes of citrus fruit, it has not been systematically examined [55].

About the physicochemical parameters of firmness, total soluble solids (TSS), and acidity, there are many reports that ethylene application does not modify them in different citrus cultivars [27, 55–57]. Nevertheless, the effect of degreening on these attributes can differ depending on the citrus cultivar. In a study that compared three early citrus cultivars, only the 'Owari' mandarin had higher softening and higher TSS in degreened fruit (2 ppm, 2–8 days) than in non-degreened ones [58].

In citrus fruit, vitamin C is widely regarded as the most important water-soluble antioxidant compound and an excellent reducing agent [59]. Vitamin C content in Citrus sp. depends on both the maturity stage and other pre- and postharvest factors. Many studies have focused on evaluating the effect of degreening treatment on vitamin C content. Sdiri et al. [6] did not observe any significant reduction in vitamin C in mandarins 'Clemenules' and 'Clemenpons' after exposing them to ethylene (2 ppm) for 48 h, 72 h, or 120 h. Mayuoni et al. [55] performed a study on different citrus cultivars and found no differences in the vitamin C content of 'Star Ruby' grapefruit and 'Satsuma' mandarins after degreening treatment lasting up to 72 h. Only 'Navel' oranges presented a slight detriment after 72 h, which was attributed to the storage period after the degreening process as no differences were found between degreened and non-degreened fruit. In orange 'Navelina' and seven early mandarins ('Basol', 'Clemenrubí', 'Clemenpons', 'Clemenules', 'Orogros', 'Oronules', 'Prenules'), no differences were observed between the degreened fruit with ethylene (2 ppm, 120 h) and the untreated fruit after cold storage to simulate quarantine conditions (1°C, 16 days), plus a shelf-life period (20°C, 7 days) [24]. In tangerine 'Batu-55', a 24-hour degreening treatment at 1, 3, or 5 ppm did not influence vitamin C content [60]. About grapefruit, no differences have been reported in 'Star Ruby' between non-degreened and degreened fruit with ethylene (2 ppm, 60 h) after a 21-day storage period at 10°C (to simulate the shipment period) and followed up to 14 days at 20°C (to simulate retail store conditions) [56]. However, in other studies ethylene has been reported to induce an increase in vitamin C. Chaudhary et al. [57] detected more vitamin C in degreened 'Rio Red' grapefruit (3.5 ppm ethylene, 72 h) after 35 storage days at 11°C than in non-degreened fruit. Sdiri et al. [61] also observed a slight increase in L-dehydroascorbic acid in degreened 'Clemenules' and 'Clemenpons' mandarins after simulating a shelf-life period (20°C, 7 days). This increased vitamin C content could be due to the dominant expression of the gene encoding L-galactose-1-phosphatephosphatase (GPP), an enzyme that is related to vitamin C biosynthesis, after finding rising GPP transcript levels in ethylene-treated tomatoes [59].

The exogenous ethylene effect can affect the enzymes involved in plant metabolic pathways, such as phenylalanine ammonia-lyase (PAL), the first enzyme in the phenylpropanoid pathway, or chalcone synthase (CHS) [62]. Regarding phenolic compounds (flavanones, flavones, polymethaoxy flavones, flavanols, hydroxybenzoic acids, and hydroxycinnamic acids), no ethylene effect has been observed in oranges 'Navel' and 'Valencia Delta', 'Batu-55' tangerine and 'Star Ruby' grapefruit when submitted to degreening [55, 56, 60, 63]. Sdiri et al. [24] studied the influence of ethylene exposure on the phenolic profile of eight early-season commercial citrus varieties during degreening treatment (2 ppm, 120 h), plus the simulation of quarantine conditions (1°C, 16 days). Only the flavanones profile was modified in the 'Clemenrubi' and 'Clemenpons' mandarins, whose fruit had been exposed to ethylene and showed the highest total flavanones content after the shelf-life period. An increase in phenolic compounds has also been reported in other studies. Higher total phenolic content has been found in the juice of lemon 'Fino' obtained from degreened fruit compared to non-degreened fruit [19]. In 'Rio Red' grapefruit, Chaudhary et al. [57] showed that degreened fruit exhibited higher limonin and flavonoids contents and lower furocoumarin levels (mainly 6–7-dihydroxybergamotin) than non-degreened fruit after 7 days of cold storage at 11°C. Nevertheless, in grapefruit 'Rio Red', no differences were detected between degreened and non-degreened fruit at the end of the study period (21 days at 11°C, plus 14 days at 21°C). Chaudhary et al. [16] evaluated the effect of ethylene concentration on the flavonoid profile in the same cultivar and found that the 10 ppm-degreened fruit had significantly higher contents for most of the phytochemicals measured compared to the 5 ppm-treated fruit.

Although exposing citrus fruit to ethylene induces carotenoid accumulation in peel [12], scarce information is available about the effect of degreening treatment on the carotenoid content in juice sacs. Matsumoto et al. [17] studied the effect of fruit exposure to different concentrations and temperatures on the carotenoid accumulation in the flavedo and juice vesicles of satsuma mandarins. The results of their study revealed that carotenoid synthesis in citrus was temperature-sensitive, and this effect was tissue-dependent. Storage at 20°C increased carotenoid accumulation in flavedo and maintained carotenoid content in juice sacs. However, storage at 5°C and 30°C slightly increased carotenoid content in flavedo and decreased contents in juice sacs. No exogenous ethylene effect on carotenoid content in the juice sacs of the fruit stored at 20°C and 5°C was observed. Chaudhary et al. [16] also reported that ethylene exposure did not affect the β-carotene and lycopene contents in 'Star Ruby' grapefruit juice during degreening treatment.

Therefore, by considering these recent findings, we conclude that degreening treatment can be used to enhance early citrus fruit peel color with minimal effects on nutritional quality.

5. Sensorial quality of degreened citrus fruit

Some studies have pointed out that degreening can impair the taste quality of citrus fruit [55, 64, 65]. Sometimes not using degreening has even been the objective to differentiate citrus fruit by assuming a superior organoleptic quality as in Protected Geographical Indication (PGI) 'Clémentine de Corse' [66]. Nevertheless, other studies report that degreening treatments performed under standard conditions do not affect sensory citrus fruit quality [6, 18, 67]. Alteration to the organoleptic quality of degreened fruit has been mainly related to low temperature during their posterior storage period [68].

There are many volatiles responsible for a flavor or aroma sensation in citrus fruit. Moreover, combinations of volatiles yield different flavors than those expected from individual compounds [69]. Citrus fruit presents a complex profile of volatile organic compounds (VOCs), and it is possible to find complex combinations of a subset of up to 300 compounds [70]. Of these, aldehydes and esters are the compounds with the strongest impact on citrus aroma. Changes in the aroma-active compounds of degreened citrus fruit have been reported as being extremely variety-dependent. The changes in the volatile profile of several citrus fruit cultivars submitted to degreening treatment, plus 16-day storage at 1°C and shelf-life, have been studied [71]. While the aroma active compounds of 'Clemenules' and 'Navelina' are not influenced by ethylene exposure, 'Oronules' and 'Clemenrubí' presented higher levels of some esters, such as ethyl propionate and ethyl octanoate, in degreened fruit than in non-degreened fruit. Other cultivars, such as 'Basol' and 'Prenules' mandarins have shown few changes caused by ethylene. It should be noted that despite the effect of degreening on certain volatile compounds in some of these varieties, no differences in sensory quality were observed. Mayuoni et al. [55] detected a minor effect of degreening treatment (4 ppm, 72 h) on volatiles content and composition in the juice of 'Navel' oranges, 'Star Ruby' grapefruit, and 'Satsuma' mandarins. While ethylene exposure did not affect the flavor of oranges and grapefruit, a slight detriment to sensory acceptability was observed in mandarins.

In recent years, several studies have reported no negative degreening effects on sensory citrus fruit properties. In 'Owari', 'Clemenules' and 'Oronules' mandarins and in 'Navelina' oranges, Morales et al. [58] have shown that exposure to ethylene (2 ppm for 2–8 days) does not bring about any significant changes in internal sensory characteristics. In fact, consumers were unable to detect the effect of ethylene degreening on physicochemical parameters, although slight differences in soluble solids and acidity were noted in some cultivars. The ethylene degreening treatment has been shown to increase consumer purchase intention. In 'Fino' lemons, a sensory analysis determined that degreened fruit subsequently stored for up 28 days at 10°C obtained a similar overall liking to non-degreened fruit. In addition, degreened lemons were perceived as having a better typical lemon aroma than non-degreened lemons [19].

6. Conclusion

We conclude that postharvest degreening treatment is a very useful tool for achieving attractive external citrus fruit color to advance the commercial season without affecting the internal quality, flavor, and nutritional properties of citrus flesh. However, it is necessary to bear in mind that this treatment must be carried out with utmost care to avoid possible physiological disorders on fruit peel. To this end, it is necessary to establish all the process parameters by taking into account the variety to be treated and the optimum harvesting time.

Author details

Julia Morales, Lourdes Cervera, Pilar Navarro and Alejandra Salvador*
Valencian Institute of Agricultural Research, Moncada, Valencia, Spain

*Address all correspondence to: salvador_ale@gva.es

References

[1] Tarancón P, Tárrega A, González M, Besada C. External quality of mandarins: Influence of fruit appearance characteristics on consumer choice. Food. 2021;**10**(9):2188. DOI: 10.3390/foods10092188

[2] Carmona L, Zacarías L, Rodrigo MJ. Stimulation of coloration and carotenoid biosynthesis during postharvest storage of 'Navelina' orange fruit at 12°C. Postharvest Biology and Technology. 2012;**74**:108-117. DOI: 10.1016/j.postharvbio.2012.06.021

[3] Lafuente MT, Alférez F, Romero P. Postharvest ethylene conditioning as a tool to reduce quality loss of stored mature sweet oranges. Postharvest Biology and Technology. 2014;**94**:104-111. DOI: 10.1016/j.postharvbio.2014.03.011

[4] Lafuente MT, Sala JM, Zacarias L. Active oxygen detoxifying enzymes and phenylalanine ammonia-lyase in the ethylene-induced chilling tolerance in citrus fruit. Journal of Agricultural and Food Chemistry. 2004;**52**(11):3606-3611. DOI: 10.1021/jf035185i

[5] González-Candelas L, Alamar S, Sánchez-Torres P, Zacarías L, Marcos JF. A transcriptomic approach highlights induction of secondary metabolism in citrus fruit in response to *Penicillium digitatum* infection. BMC Plant Biology. 2010;**10**(194). DOI: 10.1186/1471-2229-10-194

[6] Sdiri S, Navarro P, Monterde A, Benabda J, Salvador A. New degreening treatments to improve the quality of citrus fruit combining different periods with and without ethylene exposure. Postharvest Biology and Technology. 2012;**63**(1):25-32. DOI: 10.1016/j.postharvbio.2011.08.005

[7] Barry CS, Giovannoni JJ. Ethylene and fruit ripening. Journal of Plant Growth Regulation. 2007;**26**:143. DOI: 10.1007/s00344-007-9002-y

[8] Eaks IL. Respiratory response, ethylene production, and response to ethylene of citrus fruit during ontogeny. Plant Physiology. 1970;**45**:334-338

[9] Alós E, Distefano G, Rodrigo MJ, Gentile A, Zacarías L. Altered sensitivity to ethylene in 'Tardivo', a late-ripening mutant of Clementine mandarin. Physiologia Plantarum. 2013;**147**:1300-1315. DOI: 10.1111/ppl.12133

[10] Yin XR, Xie XL, Xia XJ, Yu JQ, Ferguson IB, Giovannoni JJ, et al. Involvement of an ethylene response factor in chlorophyll degradation during citrus fruit degreening. The Plant Journal. 2016;**86**(5):403-412. DOI: 10.1111/tpj.13178

[11] Fujii H, Shimada T, Sugiyama A, Nishikawa F, Endo T, Nakano M, et al. Profiling ethylene-responsive genes in mature mandarin fruit using a citrus 22 K oligoarray. Plant Science. 2007;**173**:340-348. DOI: 10.1016/j.plantsci.2007.06.006

[12] Rodrigo MJ, Zacarias L. Effect of postharvest ethylene treatment on carotenoid accumulation and the expression of carotenoid biosynthetic genes in the flavedo of orange (*Citrus sinensis* L. Osbeck) fruit. Postharvest Biology and Technology. 2007;**43**(1):14-22. DOI: 10.1016/j.postharvbio.2006.07.008

[13] Purvis AC, Barmore CR. Involvement of ethylene in chlorophyll degradation in peel of citrus fruits. Plant Physiology. 1981;**68**(4):854-856. DOI: 10.1104/pp.68.4.854

[14] Cuquerella J, Salvador A, Martínez Jávega JM, Navarro P. Effect of quarantine cold treatment on early-season Spanish mandarins. Acta Horticulturae. 2005;**682**:743-748. DOI: 10.17660/ActaHortic.2005.682.97

[15] Miller WM, Nelson B, Richard R, Ismail MA. Ethylene measurement and control in Florida citrus degreening. In: Integrated View of Fruit & Vegetable Quality. Florida: CRC Press; 2018. pp. 154-162

[16] Chaudhary PR, Bang H, Jayaprakasha GK, Patil BS. Effect of ethylene degreening on flavonoid pathway gene expression and phytochemicals in Rio red grapefruit (Citrus paradisi Macf). Phytochemistry Letters. 2017;**22**:270-279. DOI: 10.1016/j.phytol.2017.09.016

[17] Matsumoto H, Ikoma Y, Kato M, Nakajima N, Hasegawa Y. Effect of postharvest temperature and ethylene on carotenoid accumulation in the flavedo and juice sacs of Satsuma mandarin (Citrus unshiu Marc.) fruit. Journal of Agricultural and Food Chemistry. 2009;**57**(11):4724-4732. DOI: 10.1021/jf9005998

[18] Martínez-Jávega JM, Monterde A, Navarro P, Salvador A. Response of new clementines to degreening treatment. Proceedings of the International Society of Citriculture. 2008;**11**:1342-1346

[19] Serna-Escolano V, Giménez MJ, García-Pastor ME, Dobón-Suárez A, Pardo-Pina S, Zapata PJ. Effects of degreening treatment on quality and shelf-life of organic lemons. Agronomy. 2022;**12**(2):270. DOI: 10.3390/agronomy12020270

[20] Porat R, Weiss B, Cohen L, Daus A, Goren R, Droby S. Effects of ethylene and 1-methylcyclopropene on the postharvest qualities of 'Shamouti' oranges. Postharvest Biology and Technology. 1999;**15**(2):155-163. DOI: 10.1016/S0925-5214(98)00079-9

[21] Smilanick JL, Mansour MF, Sorenson D. Pre-and postharvest treatments to control green mold of citrus fruit during ethylene degreening. Plant Disease. 2006;**90**(1):89-96. DOI: 10.1094/PD-90-0089

[22] Goedhals-Gerber LL, Khumalo G. Identifying temperature breaks in the export cold chain of navel oranges: A Western Cape case. Food Control. 2020;**110**:107013. DOI: 10.1016/j.foodcont.2019.107013

[23] Zhang J, Timmer LW. Preharvest application of fungicides for postharvest disease control on early season tangerine hybrids in Florida. Crop Protection. 2007;**26**(7):886-893. DOI: 10.1016/j.cropro.2006.08.007

[24] Sdiri S, Navarro P, Monterde A, Benabda J, Salvador A. Effect of postharvest degreening followed by cold-quarantine treatment on vitamin C, phenolic compounds and antioxidant activity of early-season citrus fruit. Postharvest Biology and Technology. 2012;**65**:13-21. DOI: 10.1016/j.postharvbio.2011.10.010

[25] Van Wyk AA, Huysamer M, Barry GH. Extended low-temperature shipping adversely affects rind colour of 'Palmer Navel' sweet orange [*Citrus sinensis* (L.) Osb.] due to carotenoid degradation but can partially be mitigated by optimising post-shipping holding temperature. Postharvest Biology and Technology. 2009;**53**:109-116. DOI: 10.1016/j.postharvbio.2009.04.004

[26] Sdiri S, Navarro P, Salvador A. Postharvest application of a new growth regulator reduces calyx alterations

of citrus fruit induced by degreening treatment. Postharvest Biology and Technology. 2013;**75**:68-74. DOI: 10.1016/j.postharvbio.2012.08.004

[27] Ladaniya MS, Singh S. Use of ethylene gas for degreening of sweet orange (Citrus sinenesis Osbeck) cv. Mosambi Journal of Science and Industrial Research. 2001;**60**:662-667

[28] Cohen E. The effect of temperature and relative humidity during degreening on the coloring of shamouti orange fruit. Journal of Horticultural Science. 1978;**53**:143-146

[29] Wills RBH, Warton MA, Mussa DMDN, Chew LP. Ripening of climacteric fruits initiated at low ethylene levels. Australian Journal of Experimental Agriculture. 2001;**41**:89-92. DOI: 10.1071/EA00206

[30] Cronjé PJR, Crouch EM, Huysamer M. Postharvest calyx retention of citrus fruit. International Society of Horticultural Science. 2005;**628**:369-376

[31] Taylor JE, Whitelaw CA. Signals in abscission. The New Phytologist. 2001;**15**:323-339

[32] Alhassan N, Golding JB, Wills RBH, Bowyer MC, Pristijono P. Long term exposure to low ethylene and storage temperatures delays calyx senescence and maintains afourer mandarins and navel oranges quality. Food. 2019;**8**:19

[33] Deng L, Yin B, Yao S, Wang W, Zeng K. Postharvest application of oligochitosan and chitosan reduces calyx alterations of citrus fruit induced by ethephon degreening treatment. Journal of Agricultural and Food Chemistry. 2016;**64**:7394-7403. DOI: 10.1021/acs.jafc.6b02534

[34] Salvador A, Navarro P, Monterde A, Martínez-JávegaJM.Postharvestapplication of auxins to control calyx senescence in clementines submitted to degreening treatment. Proceedings of the International Society Citriculture. 2008;**11**:1377-1382

[35] Carvalho CP, Salvador A, Navarro P, Monterde A, Martinez-Jávega JM. Effect of auxin treatments on calyx senescence in the degreening of four mandarin cultivars. HortScience. 2008;**43**(3):747-752

[36] Martínez-Jávega JM, Monterde A, Navarro P, Salvador A. Response of new Clementines to degreening treatment. In: Proceedings of Program and Abstracts, 11th International Citrus Congress (ISC Congress); 2008; Wuhan, China, 333.

[37] Brown GE, Burns JK. Enhanced activity of abscission enzymes predisposes oranges to invasion by Diplodia natalensis during ethylene degreening. Postharvest Biology and Technology. 1998;**14**:217-227

[38] Salvador A, Navarro P, Monterde A, Martínez-JávegaJM.Postharvestapplication of auxins to control calyx senescence in clementines submitted to degreening treatment. In: Proceedings of the International Society of Citriculture; Spain. 2011. pp. 1377-1382

[39] Tormo DJ, Sdiri S, Conesa E, Navarro P, Salvador A. Application of 3,5,6-TPA under commercial conditions to control calyx senescence associated to degreening treatment. Acta Horticulturae. 2015;**1065**:1655-1661. DOI: 10.17660/ActaHortic.2015.1065.212

[40] Pássaro C, Navarro P, Salvador A. Poscosecha. In: Garcés L, Corporación Universitaria Lasallista, editors. Cítricos: cultivo, poscosecha e industrialización. 2012. pp. 223-284

[41] Krajewsky AJ. Pittaway. In: Barry GH, editor. Common Defects Associated with Degreening of Citrus. Nelspruit, South Africa: Citrus Research International (Pty) Ltd; 2002

[42] Petracek PD, Kelsey DF, Grierson W. In: Wardowski W, Miller WM, Hall DJ, Grierson W, editors. Physiological Peel Disorders. Fresh Citrus Fruit. 2nd ed. Florida, USA: Florida Science Source, Inc; 2006. pp. 397-442

[43] Albrigo LG. Water relations and citrus fruit quality. In: Sauls JW, Jackson LK, editors. Water Realations. Gainiville: University of Florida Fruit/ Crops Department; 1975. pp. 41-48

[44] Zheng Y, He S, Yi S, Zhou Z, Mao S, Zhao X, et al. Characteristics and oleocellosis sensitivity of citrus fruits. Scientia Horticulturae. 2010;**123**(3):312-317. DOI: 10.1016/j. scienta.2009.09.018

[45] Lafuente MT, Zacarias L. Postharvest physiological disorders in citrus fruit. Ingentaconnect. 2006;**9**:1-9. DOI: 10.2212/spr.2006.1.2

[46] Montero CRS, Schwarz LL, dos Santos LC, dos Santos RP, Bender RJ. Oleocellosis incidence in citrus fruit in response to mechanical injuries. Scientia Horticulturae. 2012;**134**:227-231. DOI: 10.1016/j.scienta.2011.10.026

[47] Ritenour MA, Dou H. Stem-End Rind Breakdown of Citrus Fruit: HS936/ HS193, 7/2003.EDIS, 2003, 13.

[48] Cajuste JF, González-Candelas L, Veyrat A, García-Breijo FJ, Reig-Armiñana JR, Lafuente M. Epicuticular wax content and morphology as related to ethylene and storage performance of 'Navelate' orange fruit. Postharvest Biology and Technology. 2010;**55**:29-35

[49] Ritenour MA, Miller WM, Wardowski WW. Recommendations for Degreening Florida Fresh Citrus Fruits. Circular 1170. Horticultural Sciences Department. 2003. Florida Cooperative Extension Service, IFAS, University of Florida, Gainesville.

[50] Moscoso-Ramírez PA, Palou L. Effect of ethylene degreening on the development of postharvest penicillium molds and fruit quality of early season citrus fruit. Postharvest Biology and Technology. 2014;**91**:1-8. DOI: 10.1016/j. postharvbio.2013.12.008

[51] Lorente D, Escandell-Montero P, Cubero S, Gómez-Sanchís J, Blasco J. Visible–NIR reflectance spectroscopy and manifold learning methods applied to the detection of fungal infections on citrus fruit. Journal of Food Engineering. 2015;**163**:17-24. DOI: 10.1016/j. jfoodeng.2015.04.010

[52] Wild BL. Ethylene gas burn of Washington navel oranges-a form of anthracnose induced by degreening and controlled by brushing or applying fungicides. Australian Journal of Experimental Agriculture. 1990;**30**(4):565-568. DOI: 10.1071/ EA9900565

[53] Tuset JJ, Garcia J, Hinarejos C. Effect of intermittent Degreening Method on Decay of Satsuma Mandarin. In: Goren R, Mendel K, editors. Proceeding of the Sixth International Citrus Congress. 1988. pp. 1461-1465

[54] Zhang J. *Lasiodiplodia theobromae* in Citrus Fruit (Diplodia Stem-End Rot). In: Bautista-Baños S, editors. Postharvest Decay Control Strategies. Academic Press; 2014. p. 309-335. DOI: 10.1016/ B978-0-12-411552-1.00010-7

[55] Mayuoni L, Tietel Z, Patil BS, Porat R. Does ethylene degreening

affect internal quality of citrus fruit? Postharvest Biology and Technology. 2011;**62**:50-58. DOI: 10.1016/j.postharvbio.2011.04.005

[56] Chaudhary P, Jayaprakasha GK, Porat R, Patil BS. Degreening and postharvest storage influences 'Star Ruby' grapefruit (Citrus paradisi Macf.) bioactive compounds. Food Chemistry. 2012;**135**:1667-1675. DOI: 10.1016/j.foodchem.2012.05.095

[57] Chaudhary PR, Jayaprakasha GK, Patil BS. Ethylene degreening modulates health promoting phytochemicals in Rio Red grapefruit. Food Chemistry. 2015;**188**:77-83. DOI: 10.1016/j.foodchem.2015.04.044

[58] Morales J, Tárrega A, Salvador A, Navarro P, Besada C. Impact of ethylene degreening treatment on sensory properties and consumer response to citrus fruits. Foodservice Research International. 2020;**127**:108641. DOI: 10.1016/j.foodres.2019.108641

[59] Mditshwa A, Magwaza LS, Tesfay SZ, Opara UL. Postharvest factors affecting vitamin C content of citrus fruits: A review. Scientia Horticulturae. 2017;**218**:95-104. DOI: 10.1016/j.scienta.2017.02.024

[60] Kailaku SI, Nurjanah R, Jamal IB, Broto W. The effect of degreening on antioxidants of tangerine cv. Batu-55. In: IOP Conference Series: Earth and Environmental Science. Vol. 542. 2020. p. 012018

[61] Sdiri S, Navarro P, Ben Abda J, Monterde A, Salvador A. Antioxidant activity and vitamin C are not affected by degreening treatment of clementine mandarins. Acta Horticulturae. 2012;**934**:893-899. DOI: 10.17660/ActaHortic.2012.934.119

[62] El-Kereamy A, Chervin C, Roustan JP, Cheynier V, Souquet JM, Moutounet M, et al. Exogenous ethylene stimulates the long-term expression of genes related to anthocyanin biosynthesis in grape berries. Physiologia Plantarum. 2003;**119**:175-182. DOI: 10.1034/j.1399-3054.2003.00165.x

[63] Pereira GDS, Machado FLDC, Costa J. Quality of 'Valencia Delta' orange after degreening and coating with wax. Revista Brasileira de Engenharia Agrícola e Ambiental. 2016;**20**:936-940. DOI: 10.1590/1807-1929/agriambi.v20n10p936-940

[64] Poole ND, Gray K. Quality in citrus fruit: To degreen or not degreen? British Food Journal. 2002;**104**:492-505. DOI: 10.1108/00070700210418730

[65] Baldwin EA, Bai J, Plotto A, Ritenour MA. Citrus fruit quality assessment; producer and consumer perspectives. Stewart Postharvest Rev. 2014;**10**:1-7

[66] Belmin R, Casabianca F, Meynard JM. Contribution of transition theory to the study of geographical indications. Environmental Innovation and Societal Transitions. 2018;**27**:32-47. DOI: 10.1016/j.eist.2017.10.002

[67] Tietel Z, Weiss B, Lewinsohn E, Fallik E, Porat R. Improving taste and peel color of early-season Satsuma mandarins by combining high-temperature conditioning and degreening treatments. Postharvest Biology and Technology. 2010;**57**(1):1-5. DOI: 10.1016/j.postharvbio.2010.01.015

[68] Tietel Z, Plotto A, Fallik E, Lewinsohn E, Porat R. Taste and aroma of fresh and stored mandarins. Journal of the Science of Food and Agriculture. 2011;**91**:14-23. DOI: 10.1002/jsfa. 4146

[69] Chambers E, Koppel K. Associations of volatile compounds with sensory

aroma and flavor: The complex nature of flavor. Molecules. 2013;**18**(5):4887-4905. DOI: 10.3390/molecules18054887

[70] González-Mas MC, Rambla JL, Alamar MC, Gutiérrez A, Granell A. Comparative analysis of the volatile fraction of fruit juice from different Citrus species. PLoS One. 2011;**6**(7):e22016. DOI: 10.1371/journal.pone.0022016

[71] Sdiri S, Rambla JL, Besada C, Granell A, Salvador A. Changes in the volatile profile of citrus fruit submitted to postharvest degreening treatment. Postharvest Biology and Technology. 2017;**133**:48-56. DOI: 10.1016/j.postharvbio.2017.07.001

Chapter 5

Citrus Mineral Nutrition and Health Benefits: A Review

Abduljelili Uthman and Yahaya Garba

Abstract

Citrus fruit is one of the most important fruits all over the world. Citrus fruits are essential sources of food and energy and play a critical role in supplementing healthy diets. They contain vitamins A, E, and B (thiamine, riboflavin, and niacin), minerals, and antioxidants such as flavonoids, vitamin C, phenolic compounds, and carotenoids as nutrients present in them. Mineral nutrients are essential nutrients found in many different types of plant and animal-based foods. Macro-minerals are required in large amounts while trace minerals are needed in minute quantities such as iron, zinc, and copper. Potassium is a vital nutrient in citrus fruit which regulates fruit size, fruit appearance, fruit color, and vitamin content. Fresh size and mass, percentage of rind and juice, soluble solids content (SS), titratable acidity (TA), SS/TA ratio, and industrial yield, expressed in kg of sugar per 100 kg of processed fruit or SS per box (40.8 kg) are used to evaluate fruit quality in citrus fruits. The amount of potassium below 0.4% affects tree development; otherwise, over an extensive range of variation does not generally affect plant growth. Potassium is one of the abundant elements in citrus fruits that affect both yield and quality. This chapter reviews the role of mineral nutrients in citrus production and the roles play in the human body.

Keywords: mineral nutrient, citrus fruit, Potassium, Nitrogen, macronutrients

1. Introduction

Citrus is a genus of evergreen tree or shrub that belongs to the family of Rutaceae and a native to the subtropical and tropical region of Asia [1]. There are several species of the genus citrus which include sweet orange (*Citrus sinensis*), mandarins (*C. reticulata, C. clementina, C. deliciosa, C. unshiu* and others), lemon (*C. limon*), limes (*C. aurantifolia, C. latifolia, C. limettioides, C. limetta* and others), grapefruit (*C. paradisi*), sour orange (*C. aurantium*) and shaddocks (*C. grandis*) [2, 3]. Citrus fruit is a very popular and important fruit that is cherished all over the world in the form of fruit salad, mixed salad, fruit juice, drink, and condiment in several kinds of preparations [4]. One-third of total citrus fruit production comes from orange juice production and this account for 80% of citrus fruit processing. The world production of oranges was estimated to be 76 million metric tons in 2019–2020, the largest producers were led by Brazil, China, India, the USA, Mexico, Spain, and Egypt [5].

Global citrus production reaches to 144 million metric tons in 2020. China ranking as the biggest producer of citrus followed by Brazil, India, Mexico, Spain, Egypt, Iran, and South Africa [5].

Mineral nutrients are essential for plant growth and development. They are the major class of food components necessary for the maintenance of physiological conditions that are essential for life [6, 7]. These essential nutrients are found in many different types of plant and animal-based foods. Mineral nutrients are classified into macro-minerals and trace elements. Macro-mineral is required in greater amounts and include calcium, potassium, sodium, phosphorus, magnesium, chloride, and sulfur whereas trace minerals are needed in smaller amounts, which may include iron, zinc, selenium, manganese, copper, iodine, cobalt, and fluoride. Both types of minerals support building, and maintaining healthy bones and teeth and also help to keep the muscles, heart, and brain working properly [8, 9]. Plants are the major sources of mineral elements in our diets; these substances are not synthesized in plants but are absorbed from the soil and the atmosphere. Therefore, the amount of minerals absorbed by the plants depends largely on the nutrient content and compositions of the soil where they were grown among other environmental factors [7, 10]. Additionally, nutrients play a significant role in soil fertility and make it more productive for plant growth [11]. Mineral nutrients required by citrus trees are in large quantities in order to attain adequate growth and yield, and the requirements for some of the nutrients vary with soil fertility and type.

Trace elements in fruit may be influenced by the amount of mineral composition of the soil, irrigation water, weather conditions, and the types and amounts of fertilizers used. There is a significant impact of fruit variety on mineral content [12, 13]. Trace metals in fruits are selectively accumulated and pineapple, for example, accumulated a high level of manganese as compared to other fruits studied [14]. The citrus pulp is recognized as providing some mineral elements such as potassium, calcium, phosphorus, or magnesium for human nutrition but there are other parts of the fruit that also contain these elements. The peels of orange, lime, and mandarin are not recognized in nutrition because they are non-edible components. But these peels of the fruits (orange, lime, and mandarin) like in their pulps are promising sources of mineral elements which can be used for their health properties in food products [15]. These properties can also be applied to food as a source of functional compounds [15].

2. Chemical composition of citrus fruits

Valuable natural chemicals in citrus fruits are an excellent source of different nutrients like carbohydrates, fiber, mineral elements, and vitamins required for the human body, including vitamins C, vitamins B, potassium, phosphorous, and other elements (**Table 1**). Citrus fruits contain an impressive list of other essential nutrients both glycaemic and non-glycaemic carbohydrates (sugar and fiber).

Vitamins A and E, B vitamins (thiamine, riboflavin, and niacin), minerals, and antioxidants such as flavonoids, vitamin C, phenolic compounds, and carotenoids are many nutrients present in citrus fruits. They also contain dietary fiber, which has a positive stimulatory effect on the immune system, cardiovascular, and digestive systems [17].

Component	*C. Sinensis*	*Citrus paradisi*	*C. reticulate*	*C. aurantiifolia*	*Citrus aurantium*	*C. Limon*
Moisture (g)	88.4	88.5	87.8	84.6	87.6	85
Protein (g)	0.8	1	0.9	1.5	0.7	1
Fat (g)	0.3	0.1	0.3	1	0.2	0.9
Fiber (g)	0.5	—	—	1.3	0.3	1.7
Carbohydrates (g)	9.3	10	10.6	10.9	10.9	11.1
Ash (g)	0.7	0.4	0.4	0.7	0.3	0.3
Calcium (mg)	40	30	50	90	26	70
Phosphorous	30	30	20	20	20	10
Iron (mg)	0.7	0.2	0.1	0.3	0.3	2.3
Thiamine (mg)	—	0.12	40	0.02	—	0.02 (in juice)
Riboflavin (mg)	—	0.02	—	0.03	—	0.01 (in juice)
Niacin (mg)	—	0.3	—	0.1	—	0.01(in juice)
Vitamin C (mg)	50	—	68	63	30	39 (in juice)
Carotene, μg	—	—	350	15	1104	—
Energy, K cal	43	45	—	59	48	57

Table 1.
Chemical composition of citrus fruits (per 100 g of edible portion) [16].

3. Potassium interaction with other nutrients

The positive effect between nutrients can be enhanced by the balanced application of two nutrients i.e. synergistic interaction. Antagonistic effects occur where an increase in one nutrient reduces the uptake and function of the other, thereby resulting in reduced crop quality. For example, the rate of Mg uptake can be depressed by Ca and vice versa [18]. This is due to higher affinity to Ca than to Mg in root plasma membrane binding sites [19]. The application of phosphorus is reported to reduce plant uptake and utilization of zinc [20, 21]. More so, an increase in the use of nitrogen fertilizer led to increasing in the uptake and utilization of zinc in plants [21]. Nitrogen and potassium in the fruits account for most of the nutrients removed from the soil by citrus trees each year and the interaction of K with N is considered the most important interaction. The process of converting inorganic nitrogen to organic nitrogen compounds is energy-consuming. Therefore, inorganic nitrogen absorbed by plants must be converted into amino acids and protein as much inorganic nitrogen is of no use to the plant. Good K nutrition favors the rapid turnover of inorganic nitrogen into proteins and consequently, potassium improves the effect of nitrogen fertilizer.

Mg uptake by the plant can be inhibited by the presence of high K concentrations in the soil and this may also induce Mg deficiency in plants [22]. K may also be reducing Ca uptake, where the soil is low or deficient in Ca despite the strong predominance of Ca on the exchange sites of the soils [23]. It is evident that K affects significantly the absorption and utilization of other nutrients by plants, and the appropriate K level differs in different crops.

K and N metabolism relationship has been evaluated in some studies. In contrast to the antagonistic relationship between K^+ and NH_4^+ nutrition, a positive correlation was found to exist between the acquisition rates of K^+ and NO_3^- [24, 25] and the synthesis of amino acids and proteins can be enhanced by a sufficient supply of K, which promoted N metabolism [26, 27]. Potassium (K) deficiency was found to reduce Nitrate reductase (NR), Glutamine synthetase (GS), and Glutamate synthetase (GOGAT) activities and this inhibited nitrate absorption in certain plants [28]. More so, K deficiency was reported to up-regulate the activities of GS and Glucose dehydrogenase (GDH) in Arabidopsis [29]. The metabolism of N affected by K appears to vary in different types of plants. Meanwhile, the K concentration has a significant impact on C metabolism, and the metabolic process and energy level that exist between C metabolism and N metabolism show a strong interaction [30]. K supply of 6 mM to apple dwarf rootstock seedling was optimal; as it promoted photo-assimilate transport from leaves to roots and increased nitrogen use efficiency (NUE) which influences photosynthesis. This also enhances C and N metabolizing enzyme activities, nitrate assimilation gene activities, and nitrate transport [31].

4. Effect of deficiencies of mineral elements in citrus production

When plants are potassium (K) deficient, it affects the rate of photosynthesis. Nitrogen in large quantities with a little amount of K has resulted in having drops in protein used as building blocks, disease start setting up, reduction in carbohydrates production, reduction in fruiting, increasing fruit split, creasing of fruit, and drop in plugging. A decrease in yield and low fruit quality can be as a result of a shortage of K. Negative effects of low K generally occur on fruit yield and quality before leaf deficiency symptoms. K in the leaf range of 0.5–0.8% has been observed to have decreased yield and small fruit while K concentrations of 1.2% or more produced the maximum yield of high-quality fruit [32]. No visual deficiency symptoms were observed with moderately low concentrations of K in the tree which cause a general reduction in growth [32] and production is seriously impaired when there is an onset of visual deficiency symptoms in leaves.

Mineral deficiencies in certain plant may develop into leaf chlorosis symptoms. Different mineral deficiencies in citrus can result in distinct chlorosis patterns. Mg deficiency symptoms may appear as leaf interveinal chlorosis, with chlorotic development and necrotic lesion which occurs in later stages, particularly under high light intensity [33, 34]. Citrus trees with inadequate Mg may have no symptoms in the spring growth flush, but leaf symptoms develop as the leaves age and the fruit expand and mature in the summer and fall. Magnesium deficiency symptoms occur on mature leaves following the removal of Mg to satisfy fruit requirements.

Potassium-deficient plants as observed [35] do not develop leaf chlorosis but resulted in less biomass and some changes in nitrogen metabolism. The result of the experiment also shows that magnesium deficiency produced leaf chlorosis symptoms and loss of chlorophyll in the leaf. Reduction in nitrate concentrations brought about a partial impairment of the nitrate reductase system and this is due to effects on nitrogen metabolism [35].

Calcium deficiency was the only treatment that revealed profound effects on the nitrogen economy of citrus leaves. Remarkable lower nitrogen level in leaves and reduced nitrate concentration was observed in calcium deficient [35]. Calcium deficiency with nitrogen metabolism interference resulted in an extreme reduction of the free amino acid pool. The most abundant free amino acid in citrus is proline [36] and glutamic acids, the precursor of proline synthesis were most severely affected [37]. More so, a change in the protein level of Ca-deficient leaves might be caused by the decline in the level of ribulose bisphosphate carboxylase (RuBPcase) [35]. High concentrations of NH_4^+, K^+, Ca^{2+}, Mn^{2+}, and SO_4^{2-} in the soil can induce magnesium deficiency [38]. The uptake of Mg, Ca, and other cations by the citrus plant are usually interfered with by the high concentration of potassium (K) available in the soil either due to excessive use of fertilizer or natural soil minerals [39].

5. Mineral element as a factor influencing quality of fruits in citrus production

Inadequate mineral nutrition, whether due to deficiency or excess, may generate poor fruit quality. Therefore, it is necessary to achieve nutritional balance, in order to allow plants to grow vigorously and better tolerate biotic and abiotic stresses. Fresh size and mass, percentage of rind and juice, soluble solids content (SS), titratable acidity (TA), SS/TA ratio, and industrial yield, expressed in kg of sugar per 100 kg of processed fruit or SS per box (40.8 kg) are usually used to evaluate fruit quality in citrus fruits.

Nitrogen, phosphorus, and potassium are the most important nutrients that influence fruit quality of citrus fruits. However, deficiency or excesses in other nutrients have negative effects on fruit yield and quality [40]. Nitrogen (N) increases juice content, total soluble solid (TSS) per box and per acre, and acid content. However, excessive N can induce excess vigor and promote a vegetative rather than a flowering tree, and this can result in lower yields with lower TSS per acre. In contrast, poor fruit yields are produced when there is a low N level but promoted extensive flowering. Soluble solids: acid ratio increased with phosphorus but acid content reduced [40]. Fruit production, fruit size, green fruit, and peel thickness of citrus fruits are increased with potassium. Foliar spray of potassium nitrate or monopotassium phosphate in the spring often increases the fruit size of tangerine and grapefruit, and the fruit size and total pound solids of Valencia orange [40]. The use of urea can increase flowering and fruit set by foliar application (from 6 to 8 weeks before bloom) [40].

Productivity in low-fertility tropical soils can be enhanced by the use of fertilizer and this supports adequate mineral nutrition of citrus. Nitrogen (N) and potassium (K) fertilizers to some extent produce an increase in fruit yield and quality of citrus trees [41, 42]. Potassium also affects external fruit characteristics, in such a way that as the K supply increases, fruits become larger and coarse. But, K deficiency resulted in a reduction in the number and size of fruit in all citrus varieties and the soluble solids content of juice also decrease [42].

Citrus fruit requires high mineral nutrient amounts in order to express their full growth, yield, and fruit quality potential. In some cases, soil mineral concentrations are at sufficiency levels. However, it is necessary to apply acidity correctives for nutrients to become available and be used by the plant. The main macronutrient effects observed on citrus fruit quality include; an increase in juice color intensity, soluble solid (SS) and titratable acidity (TA), rind thickness increases, and color for nitrogen (N) [43]. They also observed reduction in titratable acidity and phosphorus (P) increases SS/TA ratio; potassium decreases SS, SS/TA ratio and juice color; while Mg was reported to have slight SS, SS/TA ratio, fruit fresh mass, and size increase and reduced rind thickness [43]. However, there is a need to validate these statements as citrus fruit quality is the product of complex actions that involved several factors, acting individually or together.

6. Roles of potassium in citrus production

Potassium (K) alongside nitrogen is one of the most important nutrients for citrus production. It is needed for enzyme activation, cell division, photosynthesis, photosynthate transport, and osmoregulation. Potassium is required in more quantities in the meristematic tissues, buds, leaves, and root tips of citrus plants [44]. It plays important role in maintaining an anion-cation balance in cells, involved in protein synthesis, opening and closing of stomata, activation of enzymes, and in the turgidity of cells [45]. Potassium regulates the value of fruits through its influence on the size of the fruit, appearance of the fruit, fruit color, and vitamin contents [46].

This important nutrient plays a key role in stimulating photosynthesis, maintaining rapid root growth, and synthesis of protein from amino acids [47]. It helps to keep the stability of electric charges - an essential for ATP formation in plant chloroplasts. Stability in pH will counterbalance the insoluble and soluble macromolecular anions by the application of a large quantity of potassium. A wide range of potassium content in the leaves is required for the average vegetative growth of citrus plants [48]. The amount of K below 0.4% affects tree development; otherwise, over an extensive range of variation does not generally affect tree growth [49].

Potassium (K) is the most abundant inorganic cation and is key in ensuring optimal plant growth [50]. K is an activator of dozens of important enzymes, such as protein synthesis, sugar transport, nitrogen and carbon metabolism, and photosynthesis. It plays an important role in the formation of yield and quality improvement [51, 52]. K is also very important for cell growth, which is an important process for the function and development of plants [44]. K has strong mobility in plants, helps in regulating cell osmotic pressure, and balances the cations and anions in the cytoplasm [45, 53]. Through these processes, K is involved in the regulation of stomata opening and closing, cell elongation, and other important physiological processes.

The effects of K level on plant growth have been studied by many scientists. K applied to Red Fuji apple with 600 kg/ha K produces the highest yield and fruit quality [54]. More so, the application of 6 mM K promoted pear growth and improved photosynthetic efficiency [55]. Treatment of potassium with 500 kg/ha resulted in increased production of navel oranges with better fruit and quality parameters [56]. Potassium interacts with other nutrient ions. Magnesium (Mg) uptake in the soil is inhibited by high concentrations of K and this may lead to Mg deficiency in plants [22]. Some fruit disorders are likely to occur under low potassium conditions or high leaf N: K ratios, like plugging and creasing, and these result in less marketable fruit.

Potassium (K) has a significant role in juice acidity properties. High K concentration increases juice acidity, while a low K concentration causes a decrease in juice acidity properties. High K availability in the soil can reduce the uptake of magnesium, calcium, and ammonium for plants [42]. Nitrogen (N) and potassium (K) in the fruits are the most nutrients removed from the soil by citrus trees each year [42]. The roles played by potassium in the plant may include; root growth enhancement, drought

Flowering	Establish good early growth
Fruit set	Continued strong growth
Fruit enlargement and maturation	Maximize fruit fill and fruit size, productivity, skin quality, and vitamin C content and reduce granulation and fruit splitting
Post harvest	Maintain long-term fruit productivity

Table 2.
Potassium at citrus growth stages [58].

tolerance, decreasing water loss and wilting, improving pest and disease resistance, and reducing stalk lodging [57]. Plants required potassium in the range of 0.5 to 2% of dry matter, which is next to that of nitrogen. When the K level is in the ideal range, then a satisfactory yield of fresh fruit can be attained.

6.1 General guidelines in potassium application

Potassium is more taken up by citrus fruits than any other nutrients. Potassium application rates can be done either upon a leaf or soil analysis but soil analysis is the most commonly used one. Potassium (also called potash) is listed on the fertilizer label as K_2O and application rates of 0.8–1.4 lb. K_2O/tree are commonly used in 3–5 splits during the growing season [58]. During late fruit growth, potassium uptake usually occurred and application rates should be increased to meet this need (**Table 2**).

7. Roles of mineral nutrients in human consumption of citrus fruits

7.1 Roles of potassium and sodium in citrus fruits

Citrus fruits are good sources of potassium. An orange may provide 6% of the dietary reference intake (DRI) of potassium, while a glass of orange juice provides 10% of the DRI [59]. Potassium played an important role in regulating water, electrolyte balance, and acid–base balance in the body [59, 60]. In addition to potassium, sodium is also responsible for the regulation of water and electrolyte balance. A glass of chilled orange juice or a few fresh oranges/mandarins is very refreshing in summer and also provides the required electrolytes.

Citrus fruits contain much less amount of sodium than potassium, which is important to patients with high blood pressure regulation, a condition associated with hypertension [59]. Orange has been reported to contain more potassium than any other citrus fruit [61]. An important property of fresh citrus fruits is their low energy value, negligible sodium concentration, and their replacement by potassium, which may be important for low-fat and restricted diets. Potassium in citrus fruits enhances the normal functioning of muscles and the nervous system in the body. Without this essential mineral, the impulse to signal the body to move could not be possible. It helps in the contraction and flexing of the muscles in the heart and other organs [8].

Potassium is the most abundant mineral of citrus juices and accounts for 40 percent of the total ash content [62]. Citrus fruits are low in sodium with a value relatively low (3–4 mg/178 ml orange juice and 4.5 mg/178 ml tangerine juice) [63]. Potassium content between 4 to 6 meq is available in 100 ml of orange juice [59]. Potassium contents in orange, grapefruit, and tangerine were reported to be 237, 350, and 132 mg, respectively [64]. Potassium intake can be increased by consuming citrus fruits and juices. One medium orange and one 225 ml of a glass of orange juice provide approximately 235 and 500 mg of potassium respectively [65]. These two elements are the main cations of the cell and persons with high blood pressure are usually placed on a low-sodium diet. Although potassium deficiency in normal adults is rare, people on diuretic medicine or on an improper diet have been known to need a supplemented intake of this element. The study of the researchers had confirmed that citrus fruits (oranges and pomelos) are not a rich source of sodium [66], therefore can be used in diets for people with cardiac or kidney problems or those susceptible to osteoporosis [67–69].

7.2 Roles of calcium, magnesium, and phosphorus in citrus fruit

Calcium and magnesium are the two major divalent cations of citrus fruit. These cations are found between 8 and 15 mg/100 ml in orange, tangerine, and grapefruit juices [63]. These two mineral nutrients contributed 2–3 percent U.S. RDA per serving of 177 ml in citrus juices. Phosphorus is a bivalent cation present in the blood cells as phosphates, in protein, lipids, and carbohydrates, and in adenosine triphosphate and adenosine diphosphate (ATP and ADP). The U.S. RDA for phosphorus is 1 g; orange juice and grapefruit juice contain between 14 to 20 mg per 100 ml of phosphorus [63]. Citrus fruits compared to other fruits, such as apples, pears, melons, peaches, plums, mangoes, and bananas, are a valuable source of calcium, which plays an important role in building hard, strong bones [59].

The citric acid in orange juice may act as a chelating agent and thus increase calcium absorption by preventing the formation of insoluble salts. They are also a valuable source of phosphorus, which together with calcium, participates in the formation of strong bones and teeth [70]. Pulp from one pomelo fruit (about 600 g) provides 9 to 16% of the DRI for phosphorus while the rind (about 320 g) also provides about 30–40% less of the DRI for phosphorus than the pulp [71]. Phosphorus is important in the diet of young people, pregnant and lactating women. Plant seeds (beans, peas, cereals, and nuts) and fruits contain phytic acid (also called phytate), that is not directly available to humans [71]. Phosphorus in citrus fruit is an essential element in the building blocks of life; the ribonucleic acids (RNA), which is required for many additional biochemical and physiological processes that include energy transfer, protein metabolism, and other functions [72]. Calcium is highly implicated in the maintenance of firmness of citrus fruits [73] and its requirements in fruits are related to cell wall stability and membrane integrity [74]. Magnesium is an important primary constituent of chlorophyll and as a structural component of ribosomes, which helps in their configuration for protein synthesis [75]. It is also required for the maximum activity of almost all phosphorylating enzymes in carbohydrate metabolism.

7.3 The roles of copper, zinc, iron, and manganese in citrus fruits

Trace metals (copper, zinc, iron, manganese, etc) are needed in the body in a small or minute quantity. These plant nutrients are supplied in citrus fruit during their cultivation. They are important in many metabolic activities of the body. Iron content in one orange (200 g) can provide about 2 mg of iron. Two oranges a day can give 4 mg, which would be more than 10% of the Recommended Dietary Allowance (RDA) in the USA. The RDA is set assuming a 10% rate of intestinal absorption [76, 77].

Oranges and pomelos are the fruits richest in iron and copper, they could be recommended in cases such as hemoglobin production disorders resulting from a deficiency of these elements [61]. Pineapple contains copper, which regulates the heart rate and blood pressure [78]. Manganese in pineapple juice help to build strong bones, connecting tissues in the body and boosting the immune system [79]. A person can get rid of nausea, constipation, throat infections, and intestinal worms by consuming pineapple juice [79]. An equally important micronutrient is zinc, which protects the body against oxidative stress and stimulates immune mechanisms [80]. Its content in the peel of orange, lemon, and all grapefruit varieties was found to be significantly higher than in the pulp [61]. Cereals and vegetable diets contain phytates and these phytates inhibit zinc and calcium absorption supplied in citrus fruits therefore caution should be taken in consuming citrus fruits along with cereals and vegetables.

8. Conclusion

Citrus fruit is an evergreen tree that needs all the essential nutrients for proper metabolic functioning. A balance between macronutrients and micronutrients is needed to optimize the yield of high-quality fruit and maintain healthy trees that are tolerant to pests, diseases, and other unfavorable conditions. Citrus fruits are valuable sources of potassium, which is needed to ensure water and electrolyte balance in the body. More so, increase consumption of citrus fruits and products provide virtually all the mineral elements needed by the body to maintain good health and prevention of degenerated diseases.

Conflict of interest

The authors declare no conflict of interest.

Author details

Abduljelili Uthman[1*] and Yahaya Garba[2]

1 Department of Biochemistry, Ibrahim Badamasi Babangida University, Lapai, Nigeria

2 Department of Crop production, Ibrahim Badamasi Babangida University, Lapai, Nigeria

*Address all correspondence to: jelilibch@gmail.com

References

[1] Liu Y, Heying E, Tanumihardjo SA. History, global distribution and nutritional importance of citrus fruits. Comprehensive Reviews in Food Science and Food Safety. 2012;**11**(6):530-545

[2] Mamma D, Christakopoulos P. Biotransformation of citrus by-products into value added products. Waste and Biomass Valorization. 2014;**5**(4):529-549

[3] Waleed FA. Citrus varieties in Egypt: An impression. International Research Journal of Applied Sciences. 2019;**1**:63-66

[4] Udoh DJ, Ndon BA. Orange (Citrus spp); Crop Production Techniques for the Tropics. 2nd ed. Palmgrove, Lagos, Nigeria: Concept Publication Limited; 2016. pp. 471-479

[5] FAO Citrus Fruit Statistical Compendium. Citrus fruit, fresh and processed. Retrieved 5 February, 2022. 2020

[6] Musa A. Effect of *Craseonycteris thonglongyai* dung on some mineral elements in the leaf of *Hibiscus sabdariffa* (roselle) at market maturity. Nigerian Journal of Experimental and Applied Biology. 2017;**17**(1):85-94

[7] Musa A, Kumar N, Mamudu I, Bida II, Ibrahim S. Evaluation of the effects of *Craseonycteris thonglongyai* compost on the concentrations of some mineral elements in the leaf of *Corchorus olitorius* (jute mallow). Lapai Journal of Applied and Natural Sciences. 2016;**1**(1):19-21

[8] Nadia H. What are the Main Functions of Minerals in the Body? Newsletter; 07 December 2018

[9] Uthman A, Musa A, Amusat AO, AbdulAzeez A, Umar M, Garba Y, et al. Investigating the dietary antioxidant and mineral constituents of three commonly consumed powder vegetables sold in Lapaimarket, Lapai – Niger State, Nigeria. African Scholars Journal of Pure and Applied Science. 2021;**22**(9):247-262

[10] Aliyu HM, Morufu AI. Proximate analysis of some leafy vegetables (roselle, jute and bitter leaf). International Journal of Foods and Agricultural Research. 2006;**3**(1):194-198

[11] Toor MD, Adnan M, Rehman F, Tahir R, Saeed MS, Khan AU, et al. Nutrients and their importance in agriculture crop production: A review. Indian Journal of Pure and Applied Biosciences. 2021;**1**:1-6. DOI: 10.18782/2582-2845.8527

[12] Turra C, Fernandes EA, Bacchi MA, Barbosa Júnior F, Sarriés GA, Blumer L. Chemical elements in organic and conventional sweet oranges. Biological Trace Element Resource. 2011;**144**:1289-1294. DOI: 10.1007/s12011-011-9127-5

[13] Turra C, Fernandes EA, Bacchi MA, Tagliaferro FS, França EJ. Differences between elemental composition of orange juices and leaves from organic and conventional production systems. Journal of Radioanal Nuclear Chemist. 2006;**270**:203-208. DOI: 10.1007/s10967-006-0329-9

[14] Beattie JK, Quoc TN. Manganese in pineapple juices. Food Chemistry. 2000;**68**:37-39. DOI: 10.1016/S0308-8146(99)00149-1

[15] Barros HR, de Castro Ferreira TAP, Genovese MI. Antioxidant capacity and mineral content of pulp and peel from commercial cultivars of

citrus from Brazil. Food Chemistry. 2012;**134**:1892-1898. DOI: 10.1016/j.foodchem.2012.03.090

[16] Waleed FA. Nutritional benefit of citrus fruits. American Journal of Biomedical Science & Resources. 2019;**3**(4):AJBSR.MS.ID.000681. DOI: 10.34297/AJBSR.2019.03.000681

[17] Rafiq S, Kaul R, Sofi SA, Bashir N, Nazir F, Ahmad NG. Citrus peel as a source of functional ingredient: A review. Journal of Saudi Society of Agricultural Science. 2018;**17**:351-358. DOI: 10.1016/j.jssas. 2016.07.006

[18] Hao X, Papadopoulos AP. Effect of calcium and magnesium on growth, fruit yield and quality in a fall greenhouse tomato crop grown on rockwool. Canadian Journal of Plant Science. 2003;**83**:903-912

[19] Marschner H. Mineral Nutrition in Higher Plants. New York: Harcourt Brace; 1986

[20] Mousavi SR. Zinc in crop production and interaction with phosphorus. Australian Journal of Basic and Applied Sciences. 2011;**5**(9):1503-1509

[21] Hafeez B, Khanif YM, Saleem M. Role of zinc in plant nutrition-A review. American Journal of Experimental Agriculture. 2013;**3**(2):374-391

[22] Tränkner M, Tavakol E, Jákli B. Functioning of potassium and magnesium in photosynthesis, photosynthate translocation and photoprotection. Physiology of Plant. 2018;**163**:414-431. DOI: 10.1111/ppl.12747

[23] Morton AR, Trolove SN, Kerckhoffs LHJ. Magnesium deficiency in citrus grown in the Gisborne district of New Zealand. New Zealand Journal of Crop and Horticultural Science. 2008;**36**(3):199-213. DOI: 10.1080/01140670809510236

[24] Rufty TW, Jackson WA, Raper CD. Nitrate reduction in roots as affected by the presence of potassium and by flux of nitrate through the roots. Plant Physiology. 1981;**68**:605-609. DOI: 10.2307/4266953

[25] Coskun D, Britto DT, Kronzucker HJ. The nitrogen–potassium intersection: membranes, metabolism, and mechanism. Plant, Cell & Environment. 2016;**10**:2029-2041. DOI: 10.1111/pce.12671

[26] Ruan J, Wu X, Ye Y, Hardter R. Effect of potassium, magnesium and sulphur applied in different forms of fertilizers on free amino acid content in leaves of tea (Camellia sinensis L). Journal of the Science of Food and Agriculture. 1998;**76**:389-396. DOI: 10.1002/(SICI)1097-0010(199803)76:3<389::AID-JSFA963>3.0.CO;2-X

[27] Ruiz J, Romero L. Relationship between potassium fertilisation and nitrate assimilation in leaves and fruits of cucumber (*Cucumis sativus*) plants. The Annals of Applied Biology. 2002;**140**:241-245. DOI: 10.1111/j.1744-7348.2002.tb00177.x

[28] Hu W, Zhao W, Yang J, Oosterhuis DM, Loka DA, Zhou Z. Relationship between potassium fertilization and nitrogen metabolism in the leaf subtending the cotton (*Gossypium hirsutum* L.) boll during the boll development stage. Plant Physiology and Biochemistry. 2016b;**101**:113-123. DOI: 10.1016/j.plaphy.2016.01.019

[29] Armengaud P, Sulpice R, Miller AJ, Stitt M, Amtmann A, Gibon Y. Multilevel analysis of primary metabolism provides new insights into the role of potassium nutrition for glycolysis and nitrogen

assimilation in Arabidopsis roots. Plant Physiology. 2009;**150**:772-785. DOI: 10.1104/pp.108.133629

[30] Hu W, Coomer TD, Loka DA, Oosterhuis DM, Zhou Z. Potassium deficiency affects the carbon-nitrogen balance in cotton leaves. Plant Physiology and Biochemistry. 2017;**115**:408-417. DOI: 10.1016/j.plaphy.2017.04.005

[31] Xu X, Du X, Wang F, Sha J, Chen Q, Tian G, et al. Effects of potassium levels on plant growth, accumulation and distribution of carbon, and nitrate metabolism in apple dwarf rootstock seedlings. Frontiers in Plant Science. 2020;**11**:904. DOI: 10.3389/fpls.2020.00904

[32] Zekri M, Obreza T. Potassium (K) for Citrus Trees; UF/IFAS Extension, Gainesville, FL 32611. Publication reviewed in September, 2019. Available from: http://edis.ifas.ufl.edu/topic_series_citrus_tree_nutrients.

[33] Marschner H, Cakmak I. High light intensity enhances chlorosis and necrosis in leaves of zinc, potassium, and magnesium deficient bean (*Phaseolus vulgaris*) plants. Journal of Plant Physiology. 1989;**134**:308-315

[34] Guo W, Nazim H, Liang Z, Yang D. Magnesium deficiency in plants: An urgent problem. Crop Journal. 2016;**4**:83-91

[35] Lavon R, Salomon R, Goldschmidt EE. Effect of potassium, magnesium, and calcium deficiencies on nitrogen constituents and chloroplast components in *Citrus* leaves. Journal of American Society of Horticultural Science. 1999;**124**(2):158-162

[36] Kato T. Nitrogen metabolism and utilization in *Citrus*. Horticultural Reviews. 1986;**8**:181-216

[37] Stewart I. Nitrogen transformations in citrus trees. Proceedings of the Soil and Crop Science Society of Florida. 1961;**21**:272-282

[38] Marschner H. Mineral nutrition of higher plants. 2nd ed. Amsterdam: Academic Press; 2002. p. 889

[39] Jakobsen ST. Interaction between plant nutrients III. Antagonism between potassium, magnesium and calcium. Acta Agriculturae Scandinavica, Section B, Soil and Plant Science. 1993;**43**:1-5

[40] Zekri M. Factors affecting citrus production and quality. Citrus Industry. Dec 2011. pp. 6-9

[41] Cantarella H, Mattos D Jr, Quaggio JA, Rigolin AT. Fruit yield of Valencia sweet orange fertilize with different N sources and the loss of applied N. Nutrient Cycling in Agroecosystems. 2003;**67**:215-223

[42] Alva AK, Mattos D Jr, Paramasivam S, Patil B, Dou H, Sajwan K. Potassium management for optimizing citrus production and quality. International Journal of Fruit Science. 2006;**6**:3-43. DOI: 10.1300/J492v06n01_02

[43] Zekri M, Obreza T, Koo R. Irrigation, Nutrition, and Citrus Fruit Quality. Gainesville: Soil and Water Science Department; 2012. p. 3. Disponível em: http://itc.tamu.edu/documents/extensionpubs/Univer-sity%20of%20Florida/SL207.pdf>. Acesso em 23 fev. 2016

[44] Hepler PK, Vidali L, Cheung AY. Polarized cell growth in higher plants. Annual Review of Cell and Developmental Biology. 2001;**17**:159-187. DOI: 10.1146/annurev.cellbio.17. 1.159

[45] Hu W, Jiang N, Yang J, Meng Y, Wang Y, Chen B, et al. Potassium (K) supply affects K accumulation and photosynthetic physiology in two cotton (*Gossypiumhirsutum* L.) cultivars with different K sensitivities. Field. Crop Research. 2016a;**196**:51-63. DOI: 10.1016/j.fcr.2016.06.005

[46] Aqarab HG, Toor MD, Javid MS, Muhammad WA, Irfan H, AbuBakar I, et al. Potassium as an essential nutrient for citrus plant: A review. Acta Scientific Agriculture. 2021;**5**(7):33-37

[47] Zekri M, Obreza T. Plant nutrients for citrus trees. UF/IFAS Extension, Gainesville. Document SL 200, one of a series of the Soil and Water Science Department, UF/IFAS Extension. Original publication date December 2002. Revised February 2009. Reviewed February 2021. Visit the EDIS website at https://edis.ifas.ufl.edu

[48] Smith PF. Citrus nutrition. In: Childers NF, editor. Nutrition of Fruit Crops. New Brunswick, New Jersey: Horticultural Publications, Rutgers; 1966

[49] Rees RL, Koz RCJ. N and K fertilization effects on leaf analysis, tree size and yield of three major foliar orange cultivars. Journal of the American Society of Horticultural Science. 1975;**100**:195-198

[50] White PJ, Karley AJ. Potassium Cell Biology of Metals and Nutrients. Berlin: Springer; 2010. pp. 199-224

[51] Marschner H. Mineral Nutrition of Higher Plants. Cambridge, MA: Academic Press; 2012

[52] Oosterhuis D, Loka D, Kawakami E, Pettigrew W. The physiology of potassium in crop production. Advance Agronomy. 2014;**126**:203-234. DOI: 10.1016/B978-0-12-800132-5.00003-1

[53] Kaiser WM. Correlation between changes in photosynthetic activity and changes in total protoplast volume in leaf tissue from hygro-, meso and xerophytes under osmotic stress. Planta. 1982;**154**:538-545. DOI: 10.1007/BF00402997

[54] Jin HC, Zhang LS, Li BZ, Han MY, Liu XG. Effect of potassium on the leaf nutrition and quality of RedFuji apple. Acta Agriculturae Boreali-occidentalis Sinica. 2007;**16**:100-104. DOI: 10.3969/j.issn.1004-1389.2007.03.026

[55] Wang YZ, Zhang HP, Huang XS, Wang JZ, Cheng R, Chen GD, et al. Effect of potassium supply on plant potassium distribution and growth and leaf photosynthetic capacity of *Pyruspyrifolia*. Journal of Nanjing Agricultural University. 2017;**40**:60-67. DOI: 10.7685/jnau.201603054

[56] Lu JW, Chen F, Wan YF, Liu DB, Yu CB, Wang YQ, et al. Effect of application of potassium on the yield and quality of Navel Orange. Journal of Fruit Science. 2001;**18**:272-275

[57] Ramesh KA, Kumar N, Kavino M. Role of potassium in fruit crops-A review. Agricultural Reviews. 2006;**27**(4):284-291

[58] Available from: www.yara.us/crop-nutrition/citrus/potassium.

[59] Baghurst K. Commonwealth scientific and industrial research organization (Australia), health sciences and nutrition, horticulture Australia. In: Update of Citrus Health Benefits Literature Review. Sydney: Horticulture Australia; 2003

[60] Pohl HR, Wheeler JS, Murray HE. Sodium and potassium in health and

disease. In: Sigel A, Sigel H, Sigel RKO, editors. Interrelations between Essential Metal Ions and Human Diseases. Netherlands, Dordrecht: Springer; 2013. pp. 29-47

[61] Anna C, Ewa Z, Dmytro Y, Zvenyslava Z, Izabela G, Sylwia K. Mineral content of the pulp and peel of various citrus fruit cultivars. Biological Trace Element Research. 2020;**193**:555-563. DOI: 10.1007/s12011-019-01727-1

[62] Izquierdo L, Sendra JM. Citrus Fruits: Composition and Characterization in Encyclopedia of Food Sciences and Nutrition. 2nd ed. Cambridge, Massachusetts, USA: Academic Press, Elsevier science Ltd.; 2003

[63] Gopalan C, Sastri VPR, Balasubramaniam SC. Nutritive Value of Indian Foods. Hyderabad: National Institute of Nutrition (Indian Council of Medical Research); 1999. p. 156

[64] Gutherie H, Picciano M. Human Nutrition. St. Louis, MO: Mosby; 1995

[65] Whitney E, Rolfes S. In: Rolfes W, editor. Understanding Nutrition. 8thed. ed. Belmont, CA., USA, West/Wadsworth; 1999

[66] Paul DK, Shaha RK. Nutrients, vitamins and minerals content in common citrus fruits in the northern region of Bangladesh. Pakistan Journal of Biological Science. 2004;**7**:238-242. DOI: 10.3923/pjbs.2004.238.242

[67] Stone M, Martyn L, Weaver C. Potassium intake, bioavailability, hypertension, and glucose control. Nutrients. 2016;**8**:444. DOI: 10.3390/nu8070444

[68] Dohadwala MM, Vita JA. Grapes and cardiovascular disease. Journal of Nutrition. 2009;**139**:1788S-1793S. DOI: 10.3945/jn.109.107474

[69] Bienholz A, Reis J, Sanli P, de Groot H, Petrat F, Guberina H, et al. Citrate shows protective effects on cardiovascular and renal function in ischemia-induced acute kidney injury. BMC Nephrology. 2017;**18**(130):130. DOI: 10.1186/s12882-017-0546-1

[70] Takeda E, Taketani Y, Sawada N, Sato T, Yamamoto H. The regulation and function of phosphate in the human body. BioFactors. 2004;**21**:345-355. DOI: 10.1002/biof.552210167

[71] Otten JJ, Hellwig JP, Meyers LD. Dietary Reference Intakes: The Essential Guide to Nutrient Requirements. Washington, D.C.: National Academies Press; 2006

[72] Prabhu AS, Fageria ND, Berni RF, Rodrigues FA. Phosphorous and plant disease. In: Datnoff LE, Elmer WH, Huber DM (ed) Mineral Nutrition and Plant Disease, APS Press, StPaul, 2007; 45-55pp.

[73] Olaiya CO. Effects of three plant bioregulators on somebiochemical properties of *Lycopersicon esculentum* (L.) Mill. [Ph.D thesis], Department of Biochemistry, University of Ibadan, Nigeria. 2006

[74] Belakbir A, Ruiz JM, Romero L. Yield and fruit quality of pepper (*Capsicum annum* L.) in response to bioregulators. Horticulture Science. 1998;**33**:85-87

[75] Havlin JL, Beaton JD, Tisdale SL, Nelson WL. Soil Fertility and Fertilizers: An Introduction to Nutrient Management. 7th ed. New Jersey: Pearson Prentice Hall; 2005

[76] Ladaniya MS. Citrus Fruit Biology, Technology and Evaluation. Amsterdam: Elsevier; 2008. p. 558

[77] Gaiina Z, Daina K. Health promoting chemical components of orange juice. Proceedings of the Latvian Academy of Sciences. Section B. 2013;**67**(4/5 (685/686)):329-333. DOI: 10.2478/prolas-2013-0061

[78] Debnath P, Dey P, Chanda A, Bhakta T. A survey on pineapple and its medicinal value. Scholars Academic Journal of Pharmacy. 2012;**1**(1):24-29

[79] Hossain MF, Akhtar S, Anwar M. Nutritional value and medicinal benefits of pineapple. International Journal of Nutrition and Food Sciences. 2015;**4**(1):84-88. DOI: 10.11648/j.ijnfs.20150401.22

[80] Chasapis CT, Loutsidou AC, Spiliopoulou CA, Stefanidou ME. Zinc and human health: An update. Archives of Toxicology. 2012;**86**:521-534. DOI: 10.1007/s00204-011-0775-1

Section 2

Citrus Pests and Diseases

Chapter 6

Citrus Virus and Viroid Diseases

Faezeh Falaki

Abstract

Citrus are cultivated in a vast area worldwide and many countries grow it. Citrus fruits are delicious and everybody can eat it easily so many farmers like to grow them because of the good market. This plant has many diseases that induce various kinds of agents like fungi, bacteria, nematodes, and viruses. In this chapter, we discussed some citrus viral diseases that are very important and dangerous for fields. First, the *Citrus Tristeza virus* is explained that exists around the world. After that, you will know about other viruses like the *Citrus psorosis virus*. Viroids are another agent that causes diseases and reduces the amount of production. You learn some of them in this chapter like *Hop stunt viroid*, *Citrus exocortis viroid*, etc. The significant point of knowing citrus viral diseases is in the management of diseases. The control of viral diseases is difficult because there are no poisons or combinations to remove viruses from infected plants. If farmers or experts know about symptoms that cause viruses or viroids, they can report it to the related office and do some work to control it and it is important to the agriculture industry.

Keywords: citrus, virus, viroid, diseases, management

1. Introduction

Citrus is a genus of the Rutaceae family that is flowering, small to moderate size of trees. The genotype of citrus is available on the Ensemble website (http://www.ensembl.org) [1]. Citrus usually grows in the tropical and subtropical regions and they produce hesperidium fruit with a leathery rind. Important crops harvested from these plants are Oranges, Lemons, Grapefruits, Pomelo, and Limes. Because of their taste and benefits for humans, most countries cultivate them. The amount of citrus production around the world is 96,812,090 tons in 2020 and the amount of area harvested is 521,000 hectares in 2020 [2]. Like other biotic organisms, plants have various kinds of diseases and pests. So citrus as the same like others can be infected by viruses, nematodes, bacteria, and fungi that can impact the amount of production. One of the agents that induce loss of production is viruses. Viruses are the smallest organism in the world and they have 16–300 nm in diameter. It usually includes nucleic acids and proteins that make a particle of a virus called a virion. For a long time, they are adopted specific organisms and cells and almost all of them require alive cells or organisms for replication. Viruses are intracellular parasites because they used protein synthesis machinery (ribosomes) in their hosts. In the other words, infected cells produce viruses' proteins and other components that viruses need to infect and live. Furthermore, they achieve their energy from energy-generating metabolic pathways' host [3].

Plant viruses are the most indispensable agent that causes huge economic losses. They are widespread and almost all plant species can be infected by viruses. Characteristics of plant viruses are the same as animal viruses but their host is different. Plant viruses cannot infect animals and animal viruses cannot cause diseases in plants. Due to these features, recently, biotechnologists use plant viruses for curing some diseases like cancer in humans. Until now, plant pathologists can detect various kinds of plant viruses in plants. Because of the importance of citrus among other crops, various kinds of viruses are introduced that we mention in this chapter. In this chapter, you will read about the most important citrus viral disease (Tristeza) and after that, you know about other virus diseases that are reported. Viroids are another casual disease in plants [4]. Viroids are a subgroup of viruses. Some characteristics of viroids are similar to viruses although they have some differences like coat protein that viroids do not have and their genome is naked. Usually, their genome is circular in host cells [5–7]and the replication ways are different. They follow a rolling circle mechanism for propagation [8, 9]. The genome of viroids does not translate, so they cannot produce any proteins. The infection of viroids is occurred by interaction with the host factors in order to replicate and exert their pathogenic effects [9]. this casual disorder can cause diseases in citrus and we review it in this chapter. We will read about Exocortis and cachexia both being induced by viroids and they are a significant effect on the production amount. We try to discuss other viroids that recently they are reported by researchers. Briefly, in this chapter, you know about different viruses and viroids that are important in the citrus industry.

2. Citrus viral diseases

2 .1 *Citrus tristeza* virus

Citrus tristeza Virus (CTV) is the most indispensable virus among plant viruses because it causes the most damaging and economically important citrus disease [10]. It is called tristeza which means "sadness, melancholy" in Portuguese and Spanish [11]. The casual of this disease is a virus that has one of the largest plant RNA viruses. The virus has single-stranded, positive-sense RNA (positive polarity RNA genomes) that is placed inside a flexuous filamentous body that forms a vision of the virus. It is liked a snake. The genome of the virus is not any segment and it is monopartite. The size of the genome is approximately 19.3 k bases (kb) and it can encode 12 open reading frames (ORFs) that have various kinds of features [11, 12]. Totally, CTV has 19 proteins that each of which has a special responsibility [12]. This virus belongs to the family *Closteroviridae* and the genus *closterovirus* [11, 12]. This virus infects the member of the citrus genus like sour orange, grapefruit, lime, etc. that belongs to the Rutaceae family [11, 13, 14]. This virus infects phloem-associated cells that are responsible to transport food in the plant. So trees that are infected show a decline and yellowish leaves and as a result, infected trees die (**Figure 1**). Furthermore, it causes stem pitting in the grafting region and a reduction in fruit size and production [14]. This virus is transmitted by infected propagating material and aphids [2, 5, 8–27]. Aphids are a significant role in transmission to remote distances. Most species of aphids are *Aphis gossypii* (the melon and cotton aphid), *Aphis spiraecola* (green citrus aphid), *Toxoptera aurantii* (the black citrus aphid), and *Toxoptera citricida* (the oriental citrus aphid) [2, 28]. The last aphid is the most major aphid to transmit because it can carry the severe strain of *Citrus tristeza virus* and others cannot do it.

Figure 1.
The symptom of Citrus tristeza virus *in the infected orchard placed in the Mazandaran province, Iran. Source: Author.*

When aphids feed extract sap, virus particles penetrate to vectors' bodies and then viruses replicate in there. If aphids feed the healthy host, they can infect that tree, and virus particles enter the healthy host [28]. CTV has two strains: mild isolates cause only mild or no symptoms in sensitive citrus indicator hosts and usually result in no economic loss. Severe isolates can cause decline, stem pitting, or both and may vary in intensity [14, 19].

2.2 *Citrus psorosis virus* (CPsV)

Psorosis is another plant viral disease that is very indispensable in the infected areas. This disease is described by Swingle and Webber (1896) first and after that other researchers reported it in various kinds of citrus cultivation regions. In fact, this disease is an ancient disease among other citrus viral diseases [29]. The casual of this disease is a virus called *Citrus psorosis virus* (CPsV) [30]. This virus belongs to the genus *Ophiovirus*, the family *Ophioviridae* [21, 22, 25, 26, 31]. The virion of CPsV is kinked filaments that are 3–4 nm in diameter [29]. The shape of this virus resembles the elongated twisted and coiled [25, 26, 29].

The genome of this virus consists of three single-stranded RNAs (ssRNAs) of negative polarity [25, 26, 29, 30] So this virus is tripartite in that each RNA encodes some proteins which have various kinds of features. RNA-dependent RNA polymerase and a 24-kDa protein, which is an unknown function, are encoded by RNA 1 which has a 24-kDa weight [29–32]. Another RNA (RNA 2) is responsible to produce a movement protein and the last RNA (RNA 3) encodes the coat protein that has a 48.6 kDa weight [21, 22, 29–31]. This virus is usually transmitted by infected graft scion [29, 30] but in some regions reported some insects caused this disease to expand to remote fields. Usually, virions place in the phloem and parenchyma cells of citrus [29].

The most significant symptoms of psorosis are bark scaling in the trunk and main branches and the gum may accumulate below the bark scales that dues to impregnate the xylem and vessel occlusion and fruits bark that formed pustules on the trunks and fruits [21, 22, 29, 31]. Other symptoms that help to detect it in citrus fields are flecking, chlorotic spots, the necrotic shock of young shoots, and ringspot in leaves or fruits [21, 22, 31]. This virus does not cause an infected tree to be killed. It shows slowly decline and a loss of production [29].

This virus has two strains: strain psorosis A (PsA) and psorosis B (PsB) [2, 3, 5, 6, 8, 9, 11, 13, 14, 21–28, 30, 31, 33–44]. The symptom of psorosis A is bark-scaling in the trunk and limbs of infected field trees and staining of interior wood and leaf symptoms are flecking and spots, and shock reaction in very young shoots, in some isolates (**Figure 2**). It can also cause dieback and decrease fruit yield [21, 29, 30]. Symptoms caused by psorosis B are severe and include bark scaling even on fine twigs, gummosis, and chlorotic blotching in old leaves with pustules on the underside of leaves [21, 22, 29, 30]. and sometimes ring spots on fruits. Bark scaling often appears in 10–12-year-old trees [29].

If the strain psorosis B does not exist in any fields, you may see some leaves symptoms, and all of them are placed in the "psorosis group" diseases such as chlorotic leaf-flecking, oak-leaf patterns, ring spots, and ring patterns [29].

Totally, this virus infected citrus trees that are 10–15 years of age when they are at maximum fruit production [30] and it causes trees not to produce fruits insufficient that should have.

2.3 *Satsuma Dwarf virus* or *Citrus mosaic virus*

Both of them are shown in Japan the first and located in the *Secoviridae* family. The genome of the causal disease consists of two RNAs that are single-stranded, positive-sense. These viruses are transmitted by grafting infected budwood and mechanical with infected knives or scissors. The symptom of this agent is dwarfing because of short internodes, multiple sprouting and small leaves, and poor growth dues to the limitation of an extensive root system with a gradual decrease of yields. Sometimes, you can see various green patterns on the fruit rind of satsuma mandarin that cause a decline in commercial value which is called citrus mosaic. This virus does not make any symptoms on the trunk of infected trees. The symptom of leaves

Figure 2.
Symptoms of Citrus psorosis virus *on (a) the leaf (ring spots) and (b) bark (bark scaling), Mazandaran province, Iran. Source: Author.*

are small, boat-and spoon-shaped leaves, and the infected fruits show small in size, with a thick rind and poor taste. If trees infect with the Citrus mosaic virus, fruits have rings, mosaic, and blotches patterns on their outer surfaces. During the growth of fruit, these patterns are changed to brown and dented, resulting in a rough fruit surface. In this disease, virus particles are found in leaf cells of satsuma because they fill Intercellular spaces and tubules [33].

2.4 *Citrus variegation virus* (CVV)

This virus is another plant viral pathogenic that infects citrus cultivars. It belongs to the *Bromoviridae* family, *Ilarvirus* genus [37].

This virus is shown in the various kinds of citrus cultivars. Its symptom is variegated, mosaic, malformation on leaves, and dwarfing [2, 37]. The basis of the symptoms on citrus cultivars researchers introduced two strains: A strain can create chlorotic mottle with variable severity on the leaves and crinkles symptoms that are called infectious variegation strain. The other one is crinkly leaf strain causes distorted, puffed, or puckered leaf segments but without variegation. This agent like others transmitted by infected scion and mechanically [37].

2.5 *Citrus leaf rugose virus*

This virus has a close relationship with the *Citrus variegation virus*. Both of them place in the same genus [2]. The genome of this virus is tripartite. It means it has three single-stranded RNAs [45].

This virus induces symptoms in some cultivars like lemons, Mexican lime, and grapefruit. The rest of the citrus cultivars do not show obvious symptoms. The symptom of sensitive hosts includes leaf flecking, puckering in leaves, and stunting [2].

2.6 *Citrus leprosis Virus*

Citrus leprosis is an important disease in the North and South of the Americas and it caused millions of citrus trees to die. For the first time, it was reported from the USA. This disease damages the amount of citrus production annually in the USA. It is a very destructive disease on the Americas continent, especially on oranges and mandarins. This disease can be transmitted by the mites [13]. This disease is non-systemic. A casual disease is a heterogenic group of RNA viruses endemic to the North and South of the American continent. These viruses belong to three genera that include Cilevirus (bipartite positive sense (+) single-stranded [ss] RNA), higrevirus (tripartite (+) ssRNA), and dichorhavirus (bipartite negative sense (−) ssRNA, family Rhabdoviridae, order Mononegavirales). Cileviruses and higreviruses are active in the cytoplasm of infected plant cells but dichorhaviruses exist in the nucleus of infected plant cells. So this disease has two classifications that basis on the place of viroplasms in infected cells: CL-Cytoplasmic (CL-C) and CL-Nuclear (CL-N) [43].

The characteristic of Dichorhaviruses is short rod-like non-enveloped particles (40–50 × 100–110 nm) and encapsidate two ss (−) RNA molecules with 6 ORF. The type of virus in this genus is the orchid fleck virus (OFV). OFV has two types that include Citrus necrotic spot virus and Citrus leprosis virus Nuclear type (CL-N) [43]. The symptoms of this disease are necrotic or chlorotic spots in leaves, branches, and fruits, which progressively leads to the early drop of leaves and fruits, branch dieback, and occasionally to the death, predominantly, of the youngest citrus trees (**Figure 3**) [43].

Figure 3.
Symptom of citrus leprosis virus on (A and B) fruits, (C) twigs, (D) leaves in the United States. Source: Levy et al. [34].

The casual of the disease is Citrus leprosis virus Cytoplasmic (CL-C) which is the type species of the genus Cilevirus, family Kitaviridae. The members of this family have bacilliform or spherical virions with positive-sense single-stranded RNA genomes [20]. The symptoms of this disease are shown on twigs, leaves, and fruits that are local chlorotic and/or necrotic lesions on the surface of the host's tissue [13, 20]. If mites feed the lesion, the agent of casual disease can transmit to the healthy plant. In addition, infected shoots induced the disease in healthy plants by grafting [13]. This virus do not move systematically in its host [13, 20].

There are many ways to control disease. First, the control of mite as a vector is important. After that, infestation gardening tools like knives and scissors. The last, remove the infected twigs and control weeds [13].

2.7 *Indian citrus ringspot virus* (ICRSV)

ICRSV can infect mandarine and expand worldwide. This disease is reported by Wallace and Drake (1968) from California for the first time [46]. This virus has limitations on hosts. It attacks the mandarine. At the first, it is reported on "Kinnow" mandarin (*Citrus reticulata* cv. "Blanco," a hybrid between "King" and "Willow" mandarins) from India. This virus has a positive sense, flexuous RNA virus approximately 7.5 kb in size with six open reading frames (ORFs) belonging to the genus Mandarivirus in the Alphaflexiviridae family of order Tymovirales [42]. The shape of the virion looks like capilloviruses [47].

The symptom of this disease is bright yellow ring spots on mature leaves, followed by a rapid decline [42, 46, 54]. Several affected trees show dieback and decline type of symptoms and thus become less productive [46].

2.8 *Citrus leaf blotch virus*

This virus is placed in the Betaflexiviridae family and it is seed-borne [42]. This virus infects trifoliate rootstocks like Citrange and the symptom is bud union crease when propagated on Troyer citrange. Also sometimes it shows chlorotic blotching in young leaves. This disease detects in France for the first time. The virion of the virus is filamentous particles about 900 × 14 nm in size, with a single-stranded, positive sense, genomic RNA (gRNA) of 8747 nt, and a coat protein of about 41 kDa. The gRNA contains three open reading frames (ORFs) and untranslated regions of 73 and 541 nt at the 5′ and 3′ termini, respectively. Because this virus can be transmitted by grafting, sanitation, and quarantine of scions are very important [48].

2.9 *Bud-union crease of citrus trees*

The symptom of this disease is a dotted or continuous line of orange-yellow discoloration associated with projections or bark pegs on the inner bark surface and corresponding pits in the wood at the bud union [24, 15, 27, 49]. Sometimes, this disease causes the infected tree to decline and die or remain stunted. This disease is related to scion-rootstock combinations. Some sweet orange and lemon varieties on trifoliate orange and trifoliate orange hybrids are shown this disease. The basis on experiments, the casual of this disease is not a virus or other infection casual. It is guessed a physiological incompatibility causes this disease [49].

The infection tree shows a pale yellow color on the leaves. This symptom resembles suffering from a nutrient deficiency. Also, these trees have a poor flash that causes them to decline and die. The amount of production in infected trees is too low. In severe infection, a groove is made on the bark at the bud union. Gum-impregnated projections exist Underneath Bark strips [27].

2.10 *Citrus yellow vein clearing virus*

This disease is showed in Pakistan for the first time in 1988 on sour orange (*Citrus aurantium* L.) and lemon (*C. limon* Burm. f.) [35, 36, 50]. The symptom of this disease is strong yellow vein clearing, leaf distortion, and occasionally, ring spots and veinal necrosis [35, 36]. The symptoms of mild or moderate vein clearing or chlorosis have been observed only in young leaves during spring flushes [35]. This virus after inoculation on chenopodium showed local lesions while phaseolus Vulgaris showed systemic chlorosis, severe mosaic, blotching, and necrosis. The casual disease is the *Citrus yellow vein clearing virus* (CYVCV). The virion of the virus is the flexuous filamentous virion particle is sized between 13 and 14 nm in diameter with a modal length of 685 nm [35]. The place of virions of virus in infected plants is in the phloem [35]. This virus is a single-strand positive-sense RNA virus comprised of 7529 nucleotides (nt) [35, 44, 50] and the genome of the virus has 6 open reading frames (ORF). *Citrus yellow vein clearing virus* (CYVCV) belongs to Genus (Mandarivirus), family (Alphaflexiviridae) [50].

In China, upon becoming infected with CVYCV, the leaves of young spring or autumn shoots in lemon or sour orange trees display a water-soaked appearance and yellow, clearing veins on their ventral side. The leaves also represent chlorotic with significant crinkling and warping [23, 35].

In India, the symptoms are different and it may be shown, with mosaic and irregular ring spot-like symptoms. In drastic infection, fruits are imperfect and trees die and which changes the amount of production. This virus on herbaceous plants appears different symptoms like mosaic-like patterns, chlorosis, and necrosis on leaves. The infection in weeds is asymptomatic. Also, this virus can infect a wild grapevine and causes different symptoms that include short internodes, reduced leaf size, chlorosis, and necrosis [23].

The symptoms of the disease disappear in summer. The many possibilities are for this reaction. One of them is the high temperature causes the limitation of replication of the virus. Another reason is related to the movement of viruses. The high temperature in summer may inhibit the virus' movement in the plant, or cause the virus to transfer to other tissues in the asymptomatic tree [35].

This virus can be transmitted by grafting, knife, and scissors gardening, and insects like aphids and white flies [23, 35]. The genus of aphids can infect contain *Aphis spiraecola* Patch, *Aphis craccivora* Koch, *Aphis gossypii* Glover, and citrus whitefly *Dialeurodes citri* (Ashmead) [23, 35].

Now, this virus is expanded in many citrus-growing areas in Asia like China, Iran, and Turkey [50].

2.11 *Leaf variegation with ring spots*

The symptom of this disease is yellowish variegation on both the upper and lower surfaces of the leaves, although the yellow color is lighter on the lower surface. The shape and size of yellow areas are different but most of them are small and more or less circular in outline they have a green center and they are more similar to ring spots. Yellow spots can expand on the twigs with variegated leaves. Sometimes leaves that have yellow spots midribs are colored with gum. Fruit on branches with variegated leaves develops slight depressions or furrows, yellow in color.

The rate of transmission of this disease to healthy plants is very low. It is transmitted by grafting infected shoots. It is guessed the casual of this disease is a virus [41].

2.12 *Citrus variegation virus*

The symptom of this disease consists of leaf flecking, mosaic, malformation, and dwarfing. The casual of this disease is the *Citrus variegation virus* (CVV) which is a member of subgroup 2 of the genus *Ilarvirus* (Bromoviridae). This disease is expanding in the Mediterranean regions and the Americas continent. The symptom of disease on Citrons and lemons is more severe than on oranges and mandarine. It includes a reduction in yield and fruit malformation. This virus has two strains: infectious variegation strain, and crinkly leaf strain. The first strain can cause chlorotic mottle with variable severity on the leaves and sometimes can show crinkle symptoms on infected trees. The second strain induces distorted, puffed, or puckered leaf segments but without variegation. This disease can be transmitted by grafting and mechanical [37].

2.13 Management of citrus viral diseases

The first step for controlling viral diseases is the exclusion of the disease and forbidden to enter the infected propagation scion. The major transmission among citrus viral diseases is grafting infected buds. So, for the propagation of citrus cultivars, it is better to use certificated budwood. If viral diseases exist in a specific area, it is better to use tolerant or resistant cultivars for rootstocks. Other ways to control include cross-protection and cultural practices [2]. Furthermore, Sanitation is very important. After working with scissors and/or knives, they wash with a disinfectant solution.

3. Citrus viroid diseases

3.1 *Citrus exocortis viroid* (CEVd)

In 1948, this disease is showed with bark scaling on the rootstock of citrus trees grafted on trifoliate orange (*Poncirus trifoliata*) that is reported by Roistacher (1988) [5, 38, 51].

The agent of this disease is different. These are the smallest pathogen among the pathogenic agents of plants that are known. Usually, they have no coated protein and their body are circular. Their genome is single-stranded RNA [5, 7, 25] and they do not translate. Usually, they have between 246 and 399 or 400 nucleotides [6, 8, 9].

This viroid belongs to the *Pospiviroidae* family, the *Pospiviroid* genus. Members of this family have a rod-like or quasi-rod-like secondary structure with five structural domains (terminal left, pathogenic, central, variable, and terminal right) and a central conserved region (CCR) within the central domain (C). they do an asymmetric rolling-circle pathway for replication which occurs in the nucleus [5, 7, 8].

The significant symptoms that are caused by this viroid are bark scaling and cracking, general decline, decreased size and stunting, and leaf epinasty and cracks in the petiole (**Figure 4**) [10, 19]. This viroid can infect various kinds of citrus cultivars like *P. trifoliata* and its hybrids, rangpur lime (*Citrus limonia*), lemon (*C. limon*), and citron (*C. medica*) [10].

This agent of the disease is the same as viruses and is transmitted by grafting and infected trimming and cutting instruments [10].

3.2 Cachexia or xyloporosis (*Hop stunt viroid* (HSVd))

This disease was shown in 1950 on Parson's Special mandarin [19] and is reported by Roistacher (1983) [53]. This viroid the same as exocortis belongs to the *Pospiviroidae* family but in the classification, its genus is different and placed in the *Hostuviroid* genus. Because of a specific 6-nucleotide motif located in the Variable (V) domain of *Hop stunt viroid* that is responsible for the induction of cachexia symptoms in cachexia-sensitive hosts, this viroid was placed in the group [24, 38, 51]. In fact, those nucleotides express the symptom of *Citrus cachexia viroid* because the variable domain is responsible for pathogenicity [38]. The secondary structure of this viroid is a rod-like structure with five domains, a "central conserved region" (CCR) and a "terminal conserved hairpin" (TCH). The variable domain located at the Right of the central conserved region and basis on two strains is introduced. "pathogenic strains" that show cachexia symptoms in sensitive hosts, and "non-pathogenic strains" that infect the same sensitive hosts without inducing symptoms. Six nucleotides in this region show different symptoms. At the left of the central conserved domain is the pathogenicity domain that is responsible induce symptoms in sensitive hosts [24, 53].

The symptom of this disease includes discoloration, gumming, and browning of phloem tissue, wood pitting, and bark cracking [6, 19, 24, 38, 51, 53, 54]. If you remove the outer bark layers, you can see the discoloration of the bark. As a result, infected trees are stunted and chlorotic, and may decline and die [53]. The

Figure 4.
Symptoms of Citrus exocortis viroid *(a) decline of plant, (b) leaf epinasty as seen on grafting infected scion on Etrog citron rootstock, indicator host of CEVd, (c) bark scaling in the United States. Sources: Roy et al. [52].*

sensitive citrus cultivars are alemow (*Citrus macrophylla* Webster), clementines (*C. clementina* Hort. Ex. Tan), mandarins (*C. reticulata* Blanco), satsumas (*C. unshiu* (Macf.) Marc.), "Rangpur" lime (*C. limonia* Osb.), kumquats (*Fortunella* spp.), and hybrids like tangelos that if they can be infected, you will see the severe symptoms (**Figure 5**) [24].

3.3 *Citrus bentleaf viroid* (CBLVd) or *citrus viroid I*

This viroid belongs to Pospiviroidae family, *Apscaviroid* genus and it has 318 nucleotides. Its structure is rod-like. Because of the resemblance of a part of the

Figure 5.
Severe symptoms of cachexia, including gumming and pitting of bark and wood (A) of declining Minneola tangelo orchard from Jiroft, Kerman Province (B) in comparison with mild symptoms of cachexia in the same variety from Mazandaran Province (C). Source: Banihashemian et al. [17].

central conserved region (CCR) with the *apple scar skin viroid* (ASSVd), this viroid is located in this genus, although this viroid infected citrus cultivars. The left terminal region of *citrus bentleaf viroid* is as the same as *citrus exocortis viroid* (CEVd) [16]. The symptom of this disease is just on the leaves as curling.

3.4 *Citrus gummy bark viroid* (CGBVd) *or Citrus viroid II*

This viroid is a strain of *Hop stunt viroid* and so it places in *Hostuviroid* genus [55]. This viroid infects the sweet orange tree and causes spots or a line of reddish-brown, gum-impregnated tissue around the scion circumference especially visible near the bud union. In a severe infection, discoloration and gumming are extended and maybe to reach the main branches [39, 40, 55]. For the first time, this disease with discoloration phloem by Nour-Eldin in Egypt [39, 40]. After that, this disease is recorded in many countries such as North African and Near East countries including Saudi Arabia, Sudan, Lybia, Iran, Turkey, and Greece [40].

If this viroid infects Dortyol and Washington navel trees, they show bark scaling resembling mild psorosis scaling. It is suggested that a variant of CVd-I1 may be either the causal agent or a factor involved in gummy bark symptoms [40].

3.5 *Citrus dwarfing viroid or Citrus viroid III*

The length of this viroid genome is 294–297 nucleotides. This viroid is placed in *Apscaviroid* genus, *Pospiviroidae* family. This viroid causes citrus trees not to grow at suitable sizes and they are shorter than healthy trees [18].

3.6 *Citrus bark cracking viroid (CBCV) or Citrus viroid IV*

This viroid belongs to *Cocadviroid* genus, *Pospiviroidae* family. This genus has 246–301 nucleotides and its secondary structure is rod-like or quasi-rod-like with five domains the same as other genera. These viroids induce symptoms in citrus plants but recently this viroid causes severe symptoms on the hop. Although dues to this viroid, Etrog citron shows leaf drooping and necrosis of the leaf midribs, it does not have a minus effect on growth and yield in citrus. If CBCV and HSVd infected one tree, the symptom is severe and you will have observed yield reduction. These viroids have a synergistic effect [18, 51].

3.7 *Citrus viroid V*

This viroid like CBLVd belongs to *Pospiviroidae* family, *Apscaviroid* genus. The genome of this viroid has 293–294 nucleotides and its conformation, like others, is rodlike. This viroid induces mild but typical symptoms on citrus leaves. This viroid causes reduced tree size and yield, and on stems, you can see very small necrotic lesions and cracks, sometimes filled with gum insensitive hosts like Etrog citron [56]. As a result, Typical viroid symptoms on this host include leaf tip browning, petiole wrinkle, and browning, mid-vein necrosis, leaf epinasty, leaf bending, bark cracking, gumming, and tree stunting [7].

3.8 Management of viroid diseases

The best work is prevention and it includes (1) using a viroid-free budwood as propagation material; (2) The appropriate sanitation at cutting and trimming time; (3) treatment of cutting tools with disinfectants; and (4) adequate indexing methods [24].

4. Conclusion

Viruses and viroids have an indispensable role in plant infection. In this chapter, we discussed the most major diseases that are caused by viruses and viroids. Citrus is the most important commercial crop for many countries and people who work in fields and nurseries that cultivate citrus should know about various kinds of citrus diseases especially virus diseases because most of them are controlled easily. The prevention is the first step to manage viral diseases. Furthermore, when experts have suitable information about various kinds of symptoms, they can rapidly detect a virus disease and control it and it is a very necessary requirement for each region. **Table 1** shows a summary about citrus viruses and viroids and their detection ways.

Diseases	Causal agent	Main symptoms	Diagnostic methods	Control measures
Tristeza	*Citrus tristeza virus*	Yellowish leaves Decline	ELISA PCR	Using healthy scion The control of vector
Psorosis	*Citrus psorosis virus*	Ring spot Chlorotic spots Bark scaling	ELISA PCR	Using healthy scion The control of vector
Citrus mosaic	*Satsuma dwarf virus* or *Citrus mosaic virus*	Dwarfing Short root system	PCR	Using healthy scion
Crinkly leaf	*Citrus variegation virus* (CVV)	Variegation and mosaic on leaves Dwarfing	PCR	Using healthy scion

Diseases	Causal agent	Main symptoms	Diagnostic methods	Control measures
Citrus leaf rugose	*Citrus leaf rugose virus*	Leaf flecking Stunting	PCR	Using healthy scion
Citrus leprosis	*Citrus leprosis Virus*	Branch dieback Local chlorotic and/or necrotic lesions on leaves	PCR	Using healthy scion
Indian citrus ringspot	*Indian citrus ringspot virus*	Yellow ring spots on leaves Decline	PCR	Using healthy scion
Citrus leaf blotch	*Citrus leaf blotch virus*	Chlorotic blotching in young leaves	PCR	Using healthy scion
Bud-union crease	*Bud-union crease of citrus trees*	Bark pegs on the inner bark surface Orange-yellow discoloration underneath of bark	PCR	Using healthy scion
Citrus yellow vein clearing	*Citrus yellow vein clearing virus*	Strong yellow vein clearing, leaf distortion, veinal necrosis	PCR	Using healthy scion
Leaf variegation	*Leaf variegation with ring spots*	Yellowish variegation on leaves	PCR	Using healthy scion
Citrus variegation	*Citrus variegation virus*	Leaf flecking, mosaic, dwarfing	PCR	Using healthy scion
Exocortis	*Citrus exocortis viroid*	Bark scaling, cracking, decline	ELISA PCR	Using healthy scion
Cachexia or xyloporosis	*Hop stunt viroid*	Discoloration, bark cracking, gumming	PCR	Using healthy scion
Citrus bentleaf	*Citrus bent leaf viroid* (CBLVd) or *Citrus viroid I*		PCR	Using healthy scion
Citrus gummy bark	*Citrus gummy bark viroid (CGBVd) or Citrus viroid II*	Discoloration phloem, gumming		Using healthy scion
Citrus dwarfing	*Citrus dwarfing viroid or Citrus viroid III*	Decline	PCR	Using healthy scion
Citrus bark cracking	*Citrus bark cracking viroid(CBCV) or Citrus viroid IV*	Leaf drooping	PCR	Using healthy scion
Citrus viroid	*Citrus viroid V*	Leaf epinasty, bark cracking	PCR	Using healthy scion

Table 1.
Summary about viruses and viroids that can infect citrus trees with the way of detection and symptoms induced on citrus plants.

Author details

Faezeh Falaki
Department of Plant Protection, Faculty of Agricultural Sciences and Food Industry, Tehran, Iran

*Address all correspondence to: faezehfalaki@yahoo.com

References

[1] Terol J, Carbonell J, Alonso R, Tadeo FR, Herrero-Ortega A, Ibañez V, et al. Sequencing of 150 citrus varieties: Linking genotypes to phenotypes. Acta Horticulturae. 2015;**1065**:585-590

[2] Food and Agriculture Organization of the United Nations. FAOSTAT Statistical Database. Rome: FAO; 2022. Available from: https://www.fao.org/faostat/en/#data/QCL/visualize. [Accessed: 2022-04-22]

[3] Modrow S, Falke D, Truyen U, Schätzl H. Molecular Virology. Berlin, Heidelberg: Springer Singapore; 2013. p. 1016. DOI: 10.1007/978-3-642-20718-1_2

[4] Agrios GN. Handbook of Plant Pathology. 5th ed. California, USA: Elsevier Academic Press; 2005. p. 922. DOI: 10.1016/C2009-0-02037-6

[5] Duran-Vila N. *Citrus exocortis* viroid. In: Hadidi A, Flores R, Randles JW, Palukaitis P, editors. Handbook of Viroids and Satellites. USA: Elsevier Academic Press; 2017. pp. 169-179. DOI: 10.1016/B978-0-12-801498-1.00016-4

[6] Reanwarakorn K, Semancik JS. Correlation of *Hop stunt viroid* variants to cachexia and xyloporosis diseases of citrus. Phytopathology. 1999;**89**(7):568-574. DOI: 10.1094/PHYTO.1999.89.7.568

[7] Steyn C, Cook G, Burger JT, Maree HJ. Construction and application of infectious citrus viroids for biological indexing. Journal of Citrus Pathology. 2016. iocv_journalcitruspathology_37092. DOI: 10.5070/C431037092

[8] Eiras M, Targon MLPN, Fajardo TVM, Flores R, Itajima EW. *Citrus exocortis viroid* and *Hop stunt viroid* doubly infecting grapevines in Brazil. Fitopatologia Brasileira. 2006;**31**:440-446. DOI: 10.1590/S0100-41582006000500002

[9] Elleuch A, Marrakchi M, Lévesque D, Bessais N, Perreault JP, Fakhfakh H. Molecular variability of *Citrus exocortis* viroid in a single naturally infected citrus tree. Plant Protection Science. 2003;**39**(4):139-145. DOI: 10.17221/3833-PPS

[10] Alcántara-MendozaS,Vergara-PinedaS, García-Rubio O, Cambrón-Sandoval VH, Colmenares-Aragón D, Nava-Díaz C. Characterization of *Citrus exocortis viroid* in different conditions of indexing. Revista Mexicana de Fitopatología. 2017;**35**:284-303. DOI: 10.18781/R.MEX.FIT.1701-3

[11] Folimonova SY. *Citrus tristeza virus*: A large RNA virus with complex biology turned into a valuable tool for crop protection. PLoS Pathogens. 2020;**16**(4):e1008416. DOI: 10.1371/journal.ppat.1008416

[12] Chen AYS, Watanabe S, Yokomi R, Ng JCK. Nucleotide heterogeneity at the terminal ends of the genomes of two California *Citrus tristeza virus* strains and their complete genome sequence analysis. Virology Journal. 2018;**15**:141. DOI: 10.1186/s12985-018-1041-4

[13] Hartung JS, Guillermo León M. *Citrus leprosis* virus C (Leprosis of Citrus). Wallingford, UK: Invasive Species Compendium, CABI; 2020. DOI: 10.1079/ISC.13449.20210200730

[14] Hilf ME, Mavrodieva VA, Garnsey SM. Genetic marker analysis of a global collection of isolates of *Citrus tristeza virus*: Characterization and distribution of CTV genotypes and

association with symptoms. Virology. 2005;**95**(8):909-917. DOI: 10.1094/PHYTO-95-0909

[15] Albanese G, Ragozzino E, Davino S, Schimio R, Barba M. Survey of *Citrus tristeza virus* in Southern Italy. In: Proceedings of International Organization of Citrus Virologists (IOCV 17), International Organization of Citrus Virologists Conference Proceedings (1957-2010). 2010;**17**(17):183-187

[16] Ashulin L, Lachman O, Hadas R, Bar-Joseph M. Nucleotide sequence of a new viroid species, *Citrus bent leaf viroid* (CBLVd) isolated from grapefruit in Israel. Nucleic Acids Research. 1991;**19**(17):47-67. DOI: 10.1093/nar/19.17.4767

[17] Banihashemian SN, Bani Hashemian SM, Ashkan SM. Determination of the dominant variants of *Hop stunt viroid* in two different cachexia isolates from North and South of Iran. Journal of Agricultural Science and Technology. 2016;**18**:1431-1440

[18] Brlansky, R. Citrus viroids [Internet] 2021. Available from: https://crec.ifas.ufl.edu/media/crecifasufledu/academics/coursedocuments/Citrus_Viroids_2.pdf. [Accessed: 2022-04-22]

[19] Camps R, Castro M, Besoain X. Simultaneous detection of CTV, CEVd and HSVd using Arizona 861 S1 Citron and RT-PCR. Ciencia e Investigación Agraria. 2014;**41**(2):255-261. DOI: 10.4067/s0718-16202014000200012

[20] Chabi-Jesus C, Ramos-González PL, Postclam-Barro M, Fontenele RS, Harakava R, Bassanezi RB, et al. Molecular epidemiology of *Citrus leprosis* virus C: A new viral lineage and phylodynamic of the main viral subpopulations in the Americas. Frontiers in Microbiology. 2021;**12**:641252. DOI: 10.3389/fmicb.2021.641252

[21] De Francesco A, Costa N, García ML. *Citrus psorosis virus* coat protein-derived hairpin construct confers stable transgenic resistance in citrus against psorosis A and B syndromes. Transgenic Research. 2017;**26**:225-235. DOI: 10.1007/s11248-016-0001-2

[22] De Francesco A, Costa N, Plata MI, García ML. Improved detection of *Citrus psorosis virus* and coat protein-derived transgenes in citrus plants: Comparison between RT-qPCR and TAS-ELISA. Journal of Phytopathology. 2015;**163**: 915-925. DOI: 10.1111/jph.12392

[23] Dey K, McVay J. Citrus Yellow Vein Clearing Virus. Florida Department of Agriculture and Consumer Services, Division of Plant Industry; 2022

[24] Eiras M, Silva SR, Stuchi ES, Carvalho SA, Garcêz RM. Identifcation and characterization of viroids in 'Navelina ISA 315' sweet orange. Tropical Plant Pathology. 2013;**38**(1):58-62. DOI: 10.1590/S1982-56762013000100009

[25] Falaki F, Alavi SV, Rakhshanderoo F. *Citrus psorosis virus*, causal agent of ring pattern disorder in Thomson Navel trees in east of Mazandaran. Iranian Journal of Applied Entomology and Phytopathology. 2013;**80**(2):161-172 (In Persian with English summary)

[26] Falaki F, Alavi SV, Rakhshanderoo F, Mousavi A. Partial biological and molecular characterization of *Citrus psorosis virus* infecting Satsuma mandarin (*Citrus unshiu* Marc) in the east of Mazandaran province. Iranian Journal of Applied Entomology and Phytopathology. 2019;**87**(2):195-207 (In Persian with English summary)

[27] Fernandez Valiela MV, Fortugno C, Corizzi F. Incidence of bud-union crease in citrus trees grafted on trifoliata rootstock in the delta del Paranti and san

Pedro areas of Argentina. International Organization of Citrus Virologists Conference Proceedings (1957-2010). 2011;**3**:3. DOI: 10.5070/C56cv8f38v

[28] Komazaki, S. Biology and virus transmission of *Citrus aphids* [Internet]. 2011. Available from: https://www.fftc.org.tw/htmlarea_file/library/20110712190621/tb136.pdf. [Accessed: 2022-04-22]

[29] Achachi A, Barka EA, Ibriz M. Recent advances in *Citrus psorosis virus*. Virus Disease. 2014;**25**(3):261-276. DOI: 10.1007/s13337-014-0199-7

[30] Reyes CA, Zanek MC, Velazquezquez K, Costa N, Plata MI, García ML. Generation of sweet orange transgenic lines and evaluation of *Citrus psorosis virus*-derived resistance against psorosis A and psorosis B. Journal of Phytopathology. 2011;**159**:531-537. DOI: 10.1111/j.1439-0434

[31] Reyes CA, Peña EJ, Zanek MC, Sanchez DV, Grau O, García ML. Differential resistance to *Citrus psorosis virus* in transgenic *Nicotiana benthamiana* plants expressing hairpin RNA derived from the coat protein and 54K protein genes. Plant Cell Reports. 2009;**28**:1817-1825. DOI: 10.1007/s00299-009-0781-4

[32] Reyes CA, Ocolotobiche EE, Marmisollé FE, Luna GR, Borniego MB, Bazzini AA, et al. *Citrus psorosis virus* 24K protein interacts with citrus miRNA precursors, affects their processing and subsequent miRNA accumulation and target expression. Molecular Plant Pathology. 2016;**17**(3):317-329. DOI: 10.1111/mpp.12282

[33] Koizum, I.M.; Iwanami, T. Satsuma dwarf [Internet]. 2008, Available from: https://iocv.ucr.edu/citrusdiseases/satsumadwarf#:~:text=Citrus%20mosaic%20virus%20(CiMV)%2C,of%20new%20leaves%20of%20C. [Accessed: 2022-04-22]

[34] Levy A, Batuman O, Sieburth P, Diepenbrock L. *Citrus leprosis* poses a threat to Florida. Citrus Industry. 10-12 Feb 2020

[35] Liu C, Liu H, Hurst J, Timko MP, Zhou C. Recent advances on citrus yellow vein clearing virus in citrus. Horticultural Plant Journal. 2020;**6**(4):216-222. DOI: 10.1016/j.hpj.2020.05.001

[36] Loconsole G, Önelge N, Potere O, Giampetruzzi A, Bozan O, Satar S, et al. Identification and characterization of citrus yellow vein clearing virus, a putative new member of the genus Mandarivirus. Phytopathology. 2012;**102**:1168-1175

[37] Moreira L, Albertazzi FJ, Garita L, Ortiz B, Villalobos W. First report of *Citrus variegation virus* in Palestine sweet lime, as coffee shade in Costa Rica. In: Proceedings of International Organization of Citrus Virologists (IOCV 18). 2010. São Paulo, Brazil: IEEE; 2011

[38] Murcia N, Bani Hashemian SM, Serra P, Pina JA, Duran-Vila N. Citrus viroids: Symptom expression and performance of Washington navel sweet orange trees grafted on Carrizo citrange. Plant Disease. 2015;**99**:125-136. DOI: 10.1094/PDIS-05-14-0457-RE

[39] Nelge NO, Cinar A, Szychowski JA, Vidalakis G, Semancik JS. Citrus viroid II variants associated with 'Gummy Bark' disease. European Journal of Plant Pathology. 2004;**110**:1047-1052

[40] Nelge N, Smar A, Kersting U, Semancik JS. Viroids associated with citrus gummy bark disease of sweet orange in Turkey. Thirteenth IOCV Conference, 1996-Viroids. International Organization of Citrus Virologists

Conference Proceedings (1957-2010). 1996;**13**(13):245-248

[41] Planes S, Mart F. Leaf variegation with ring spots. International Organization of Citrus Virologists Conference Proceedings (1957-2010). 2011;**5**:5. DOI: 10.5070/C52m70156m

[42] Prabha K, Baranwal VK. Indian citrus ringspot virus: Localization of virus in seed tissues and evidence for lack of seed transmission. Phytoparasitica. 2011;**39**:491-496. DOI: 10.1007/s12600-011-0181-5

[43] Ramos-González PL, Chabi-Jesus C, Guerra-Peraza O, Tassi AD, Kitajima EW, Harakava R, et al. *Citrus leprosis* virus N: A new dichorhavirus causing *Citrus leprosis* disease. Phytopathology. 2017;**107**(8):963-976. DOI: 10.1094/PHYTO-02-17-0042-R

[44] Rehman AU, Li Z, Yang Z, Waqas M, Wang G, Xu W, et al. The coat protein of citrus yellow vein clearing virus interacts with viral movement proteins and serves as an RNA silencing suppressor. Viruses. 2019;**11**:329. DOI: 10.3390/v11040329

[45] Scott SW, Ge X. The nucleotide sequence of *Citrus leaf rugose virus* RNA 1. The Journal of General Virology. 1995;**76**:3233-3238. DOI: 10.1099/0022-1317-76-12-3233

[46] Sharma S, Singh B, Rani G, Zaidi AA, Hallan VK, Nagpal AK, et al. In vitro production of *Indian citrus ringspot virus* (ICRSV) free Kinnow plants employing thermotherapy coupled with shoot tip grafting. Plant Cell Tissue and Organ Culture. 2008;**92**:85-92. DOI: 10.1007/s11240-007-9307-3

[47] Rustici G, Accotto GP, Noris E, Masenga V, Luisoni E, Milne RG. Indian citrus ringspot virus: a proposed new species with some affinities to potex-, carla-, fovea- and allexiviruses. Archives of Virology. 2000;**145**:1895-1908

[48] Vives MC, Galipienso L, Navarro L, Moreno P, Guerri J. Citrus leaf blotch virus: A new citrus virus associated with bud union crease on trifoliate rootstocks. Fifteenth IOCV Conference. Egypt, Cyprus: Other Viruses. 2001

[49] Salibe AA. Studies on bud-union crease of citrus trees. International Organization of Citrus Virologists Conference Proceedings (1957-2010). 2011;**3**:3. DOI: 10.5070/C59q7036bm

[50] Zhang YH, Liu CH, Wang Q, Wang YL, Zhou CY, Zhou Y. Identification of *Dialeurodes citri* as a vector of citrus yellow vein clearing virus in China. Plant Diseases. 2018;**103**(1):65-68. DOI: 10.1094/PDIS-05-18-0911-RE

[51] Vernière C, Perrier X, Dubois C, Dubois A, Botella L, Chabrier C, et al. Citrus viroids: Symptom expression and effect on vegetative growth and yield of clementine trees grafted on trifoliate orange. Plant Disease. 2004;**88**:1189-1197. DOI: 10.1094/PDIS.2004.88.11.1189

[52] Roy A, Kumar A, Walia Y, Hallan V, Ramachandran P. Studies on viroids occurring in India. In: Mandal B, Rao GP, Baranwal VK, Jain RK, editors. A Century of Plant Virology in India. Singapore: Springer; 2017. pp. 487-511. DOI: 10.1007/978-981-10-5672-7_22

[53] Serra P, Gago S, Duran-Vila N. A single nucleotide change in *Hop stunt viroid* modulates citrus cachexia symptoms. Virus Research. 2008;**138**:130-134. DOI: 10.1016/j.virusres.2008.08.003

[54] Semancik JS, Roistacher CN, Duran-Vila N. A new viroid is the causal agent of the *Citrus cachexia* disease. International Organization of Citrus

Virologists Conference Proceedings (1957-2010). 2011;**18**(18):125-135

[55] Wikipedia Website [Internet]. Citrus gummy bark viroid; 2021. Available from: https://en.wikipedia.org/wiki/Citrus_gummy_bark_viroid. [Accessed: 2022-09-14]

[56] Serra P, Eiras M, Bani-Hashemian SM, Murcia N, Kitajima EW, Daròs JA, et al. Citrus viroid V: Occurrence, host range, diagnosis, and identification of new variants. Virology. 2008;**98**(11):1199-1204. DOI: 10.1094/PHYTO-98-11-1199

Chapter 7

Quorum Quenching Bacteria: An Approach for Phytopathogens Control in Citrus Cultivars

Juan Carlos Caicedo and Sonia Villamizar

Abstract

Cell-to-cell communication system quorum sensing (QS) denotes the ability of bacteria to track the population density, in order to coordinate its phenotypic traits to successfully establish and thrive in new ecological niches. Different citrus phytopathogenic bacteria such as: *Xanthomonas citri spp. citri*, *Xillela fastidiosa* and *Pseudomonas syringae* pv. *syringae* regulate several pathogenicity factors through well-established quorum sensing DSF (Diffusible Signal Factor) and AHL (AcylHomoserine Lactone) pathways. The goal of this chapter is to review exophytic and endophytic bacteria able to disrupt quorum sensing communication system in these bacteria in order to reduce the symptomatology of citrus canker, citrus variegated chlorosis and citrus blast. The quorum quenching of phytopathogen bacteria could afford new tools for disease control, thus reducing the overuse of antimicrobial drug and decrease its environmental accumulation, thus relieving the selection pressure of resistant bacterial populations.

Keywords: citrus canker, quorum sensing, auto-inducer, biological control, biofilm

1. Introduction

Nowadays, at the bacterial world, it is widely accepted that there are biological processes that must be pointed by a coordinated behavior of entire population. Factors such as: virulence factors production, biofilm formation, secondary metabolite production, and bioluminescence are fruitless when undertaken by a single bacterium proceeding isolated [1]. A wide variety of bacteria are endowed with encoding genes for components of a cell-to-cell communication system termed as quorum sensing (QS). This QS system enables bacteria to regulate their behavior in a cell density fashion in order to modulate a gene set that enables the bacteria to adapt to environmental challenges [2]. Quorum sensing relays its activity in a production, releasing, and perception of small signal molecules called auto-inducers. In Gram-negative bacteria, the usual auto-inducers are small molecules, i.e., acyl lactone and short-chain fatty acids. The cognate receptor involved in perception of these auto-inducer molecules is: cytoplasmic transcription factor and two components histidine sensor kinases. The complex produced by the auto-inducer and receptor leads the promotion of target genes regulated by quorum sensing [3]. Gram-positive bacteria use mainly short peptides as auto-inducers, and its related receptors are transmembrane histidine sensor kinase. Usually, the union of auto-inducer and receptor triggers expression of encoding gene for AI (auto-inducer) synthase, which increases the extracellular AI concentration switching on the bacteria quorum sensing mode [4].

Disruption of quorum sensing communication system, which is termed quorum quenching, leads to a reduction in virulence factors expression without compromising bacterial survival [5]. Since a wide diversity of bacterial cells that use QS display a significant competitive advantage over other prokaryotes and eukaryotes with which they coexist in the same ecological niche, it is rational that the contender microorganisms have developed mechanism to disrupt the QS communication systems present in the ecological niche. Interference with the quorum sensing communication system either by natural or synthetic approaches may afford strategies for disease control, by reducing the virulence or turn the pathogens more susceptible to antibiotic therapy. The design, development, and employment of these approaches will depend in great measure upon the knowledge of mechanistic details of quorum sensing pathway such as: auto-inducer synthesis, signal perception, signal transduction, and genes under quorum sensing regulation [6].

Citrus is the most commercialized horticultural product in the world; however, farmers in the last two decades have seen production reduced by average of 65%, due to devastating bacterial diseases such as: Bacterial Citrus Canker (BCC) caused by *Xanthomonas citri* subsp. *citri* (Xcc), Citrus Variegated Chlorosis (CVC) caused by *Xylella fastidiosa*, Citrus Blast caused by *Pseudomonas syringae* pv. *Syringae*, and Citrus Greening or Huanglongbing (HLB) caused by *Candidatus liberibacter* sp. All bacteria aforementioned except the *Candidatus liberibacter* are endowed with quorum sensing systems, which are responsible for the pathogenesis and symptomatology in citrus host. The main objective of this chapter is to describe quorum sensing pathways in these phytopathogen bacteria, as well to review some successful approaches based in quorum sensing disruption in order to decrease disease severity.

2. Citrus canker and DSF quorum sensing pathway in *X. citri* subsp. *citri*

Bacterial citrus canker (BCC) is one of the major citrus diseases, almost all varieties of citrus crops are affected, and the severity of disease depends on bacterial species and weather conditions [7]. The etiological agent of BCC is the Gram-negative bacterium *X. citri* subsp. *citri* [8]. Nowadays, three types of BCC are recognized, which are: (i) citrus canker type A, also known as Asian citrus canker, is the most widespread disease. The BCC has a pronounced host range producing symptoms in: *Chrysopelea paradisi*, *C. aurantifolia*, *C. sinensis*, and *C. reticulata* (ii) Citrus canker type B is caused by the bacterium *Xanthomonas fuscans* subsp. *aurantifolii* type B (*XauB*) [9]. The symptomatology development is similar to citrus canker type A; however, because of the *XauB* slower growth rate, the symptoms spent more time to appear. Host range is limited to *C. limon*; however, *XauB* was rarely isolated from *C. sinensis* and *C. paradisi* [10]. Citrus canker type C is produced by *X. fuscans* subsp. *aurantifolii* type C (*XauC*). The symptomatology is similar to citrus canker type A; nevertheless, its host range is restricted to *C. aurantifolia* [9].

Pathognomonic symptoms of BCC type A are the raised corky and spongy lessons surrounded by a water-soaked margin, which are present in leaves and fruits. This lesion results from the hypertrophy and hyperplasia of mesophilic cells. This cell division disorder is induced by the bacterial effector from family AvrBs3/PthA [11].

The bacterium *Xcc* is outfitted with a vast arsenal of organelles responsible for the pathogenic traits in citrus host. The main known are: bacterial attachment, antagonism, effector production, quorum sensing regulation, and biofilm formation. For an in-depth review, please refer to Caicedo and Villamizar [12].

2.1 DSF quorum sensing pathway in Xcc

Xcc bacteria have a quorum sensing system whose auto-inducer molecule (AI) is a short-chain fatty acid belonging to DSF (diffusible signal factor) family. The DSF auto-inducer family modulates the expression of virulence and pathogenicity in several pathogenic bacteria to plants and humans [13]. The DSF molecules display a *cys* unsaturated double bond at position two as a distinguished feature in the family. The DSF family are cis-2-unsaturated fatty acids (**Figure 1**), the cis-11-methyl-2-dodecenoic acid was the auto-inducer molecule characterized to be responsible for the signaling processes in *Xcc* [15].

The discovery of the DSF signaling molecule came within a genomic study that seeks to identify a gene cluster termed *rpf* (regulation of pathogenicity factor) *rpfb-* in the bacterium *Xanthomonas campestris* pv. *campestris*. Researchers found that mutation of components of this gene cluster drives the decrease in extracellular enzyme production and exopolysaccharide as well as reduction in pathogenicity in plant susceptible [16]. Later works established the participation of these genes in the coding of elements belonging to quorum sensing communication system involved in the synthesis and perception of the DSF signal molecule [17]. The gen *rpf*F encodes an enzyme, amino acids sequence of which is related to enoyl CoA hydratase; this enzyme is responsible for DSF synthesis. The gen *rpf*B encodes and Acyl CoA ligase, which participates to a lesser extent in the synthesis of DSF auto-inducer [16].

DSF perception and signal transduction are encoded by an *rpfGHC* gene operon. *rpfC* encodes the receptor RpfC protein. This protein has a transmembrane domain, which is involved in the perception of DSF auto-inducer and the cytoplasmic domains: His-Kinase A (phosphoacceptor), His-Kinase-like ATPase, REC domain (receiver

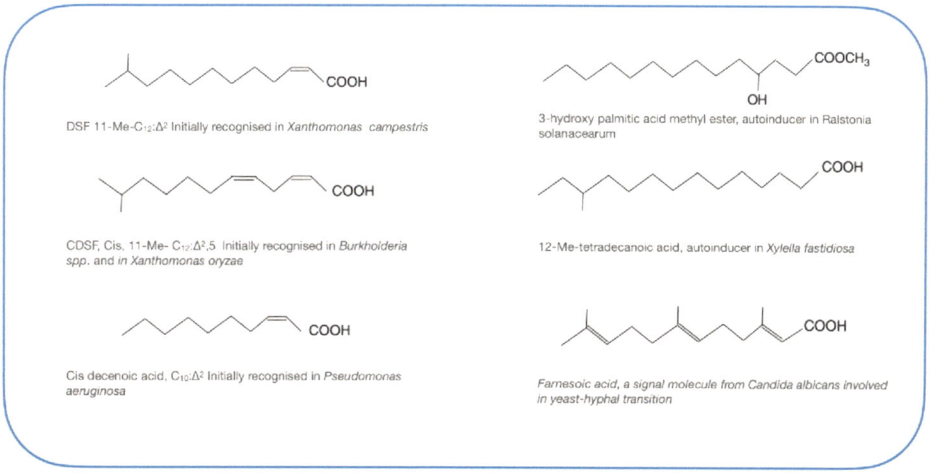

Figure 1.
(Left) Structure of the main molecules from DSF auto-inducers family and some related putative auto-inducers. (Right) Related molecules with signaling activity. Notice that these molecules lack the cys unsaturated double bond at position two. (Adapted from Ryan and Dow [14]).

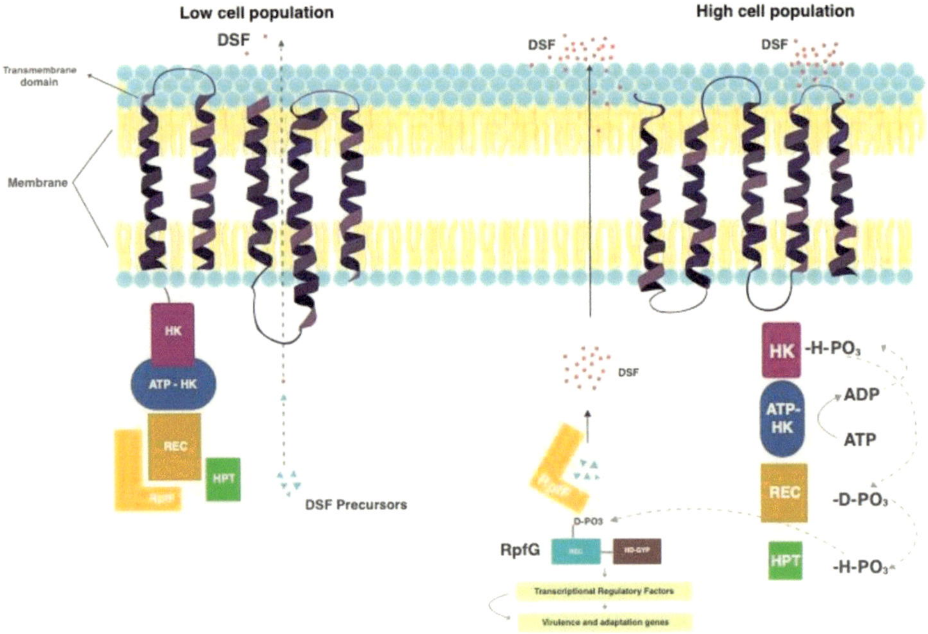

Figure 2.
At low cell population condition, RpfF the DSF synthase remains bound to REC domain from RpfC sensor, which maintains compact conformation. At high cell population condition, the DSF binds to RpfC sensor domain and induces a conformational change releasing the synthase RpfF, thus triggering an autophosphorylation and phosphorelay and the subsequent phosphotransfer to the REC domain of response regulator RpfG. The phosphorylation of RpfG, it triggers the activation of RpfG as a cyclic di-GMP phosphodiesterase reducing the level of cyclic di-GMP and releasing Clp that promotes the synthesis of extracellular enzymes and EPS (take it from Caicedo et al. [18]).

domain), and finally, the HTP domain histidine phosphotransfer (**Figure 2**). The gene *rpfG* encodes for response regulator protein, this protein include two domains: (i) an REC domain, which receives the phosphate of HTP domain from the RpfC; and (ii) the HD- GYP, which displays phosphodiesterase activity responsible for degradation of second messenger cyclic di-GMP. At physiological levels of cyclic di-GMP, the transcriptional activator cAMP-receptor-like-protein Clp remains bound to second messenger cyclic di-GMP. Consequently, reduction of cyclic di-GMP via DSF in *Xcc* leads to the release of Clp, thus allowing Clp to activate the expression of several gene and proteins involved in virulence such as: extracellular enzyme production, EPS production, biofilm formation, motility, iron uptake [19]. Furthermore, DSF signaling also positively controls the expression of *clp* gene, which suggests a supplementary regulatory complexity [14]. Summarizing, the perception of auto-inducer molecule DSF in *Xcc* by the receptor RpfC triggers the expression of its virulome. For all aforementioned, the disruption of this quorum sensing communication system could turn into a valuable tool for reducing citrus canker severity.

3. Variegated chlorosis, *Xylella fastidiosa and* quorum sensing

CVC (Citrus Variegated Chlorosis) is a disease that affects sweet orange (*Citrus sinensis*) and grapefruit (Citrus x paradise); no other citrus are susceptible to disease

[20]. CVC is caused by the Gram-negative bacterium *X. fastidiosa (Xf)*. Transmission paths of CVC are: (i) *Propagative vegetative material*: CVC is a paradigm of accidental long-distance propagation of phytopathogen; this is mainly due to the movement made by agriculture workers of infected plant material [21]. (ii) *Insect vector borne:* vectors of *Xf* are xylem sap sucking insects belonging to the Hemiptera order including sharpshooter, leafhoppers, and spittlebugs [22].

CVC symptomatology in susceptible host sweet orange initiates as a small chlorosis, which extents irregularly on the upper surface of mature leaves, the affected leaves display a consequent brownish gum-like material on the lower surface. At this disease stage, the lesions are present only in one or two branches. At the later disease stage, bacteria became systemic spreading in the plant canopy, symptoms become apparent between 3 and 6 months [22].

The bacterium *X. fastidiosa* belongs to the gammaproteobacteria group, *Xanthomonadaceae* family. *Xf* is a xylem-limited bacterium, which is obligatory colonizer of plants and insect vectors [21]. There are three major monophyletic subspecies of *X. fastidiosa: Xf* subsp. *multiplex, Xf* subsp. *Fastidiosa,* and *Xf* subsp. *pauca*, endemic respectively in North, Central, and South America [23–25]. Nowadays, it is widely accepted that *Xf* has a commensal relationship with a huge number of plant species; however, just a few numbers of clades and specific bacterial genotypes are associated as phytopathogens. Contrasting with the relationship between *X. fastidiosa* and the plant, the association of *X. fastidiosa* with insect vector is independent of plant-pathogen mixtures. Vectorial transmission is the only natural mechanism spread. There exist two principal xylem-sap feeders insect vector groups: the sharpshooter leafhoppers (Cicadellidae subfamily Cicadellinae) and spittlebugs (Cercopoidea, families Aphrophoridae, Cercopidae, and Clastopteridae) [26, 27]. Actually it is extensively known that all insect vectors aforementioned display the ability to transmit all genotypes of *X. fastidiosa* without any specificity [28]. *Xyllela fastidiosa* colonize plants gradually, the initial stage is a vessel obstruction because of bacterial multiplication, reduction of sap flow in the xylem system is due to plant response (i.e., tylose) and bacterial dissemination between vessel via pit membrane [29].

3.1 Quorum sensing in *Xyllela fastidiosa*

As mention before, the virulence of *X. fastidiosa* is related with its capability to travel and to multiply within xylem vessels, and symptoms could basically be an unintended effect caused by effective colonization that restricts with xylem sap flow. *X. fastidiosa* as similar to related phyto-pathogenes Xanthomonas and Stenotrophomonas use small molecules from DSF family in order to coordinate its behavior in a cell-density-dependent fashion [30]. Cell-to-cell communication system in *X. fastidiosa* involving the production of DSF *rpfF* gene is responsible for the DSF production as same of Xanthomonas bacteria. Unlike, *Xanthomonas* bacteria, in which the disruption of DSF quorum sensing pathway reduces its virulence and pathogenicity in *X. fastidiosa,* mutants of *rpfF* and subsequently deficient in DSF production show an hypervirulent phenotype behavior in a susceptible host such as sweet orange and grapefruit [31]. In *X. fastidiosa,* the DSF molecule is 12 methyl tetradecanoic acid *xf*DSF (**Figure 1**). The whole mechanistic details in DSF pathway are not yet completely understood. Previous studies have proposed the existence of two different types of receptors in *X. fastidiosa:* the first one RpfC transmembrane has a high sequence similarity with the RpfC of *Xanthomonas campestris*, the only difference lies that in the *X. fastidiosa* is truncated at the N terminus, apparently its function is to bind DSF and in that way to execute a negative feedback regulation in DSF production [32]. Another potential DSF sensor has an intracytoplasmic localization, and its function could be the perception of DSF accumulated within the cell. Once DSF bound to intracellular DSF receptor the autophosphorylation and phosphorelay to a response regulator as RPFG is triggered. It allows the expression of genes involved in attachment and biofilm formation [32].

Previous studies have shown that *rpfF* mutants of *X. fastidiosa* display a hypervirulent phenotype in a susceptible plant host, and this mutant strain was incapable to colonize and be spread by insect vectors. These observations arose the hypothesis that DSF signaling is used as a lifestyle dependent switch, because *rpfF* promotes the genes expression involved in attachment and biofilm formation in the xylem vessels. By contrast, *rpfF* in *X. fastidiosa* represses the expression of genes intricated in motility and hydrolytic enzyme production, which are responsible for the cell migration and pit membrane disruption. *X. fastidiosa* xylem vessels attached cells display reduced pathogenicity and a phenotype highly favorable to be acquired by the insect vector [33]. On the other hand, *rpfC* deletion mutants displayed an avirulent phenotype, because the great DSF production, these mutants were capable to successfully colonize the insect vector, whoever these bacterial cells display an impaired ability to be transferred to another susceptible plants. Finally, similar to Xanthomonas bacteria, DSF-dependent signaling regulates decyclic di-GMP in *X. fastidiosa* [32].

4. Citrus blast, black pit, *Pseudomonas syringae* pv. *Syringae,* and quorum sensing

Citrus blast is an important bacterial disease that affects commercially important citrus fruits such as sweet orange (*C. sinensis*) and mandarin (*Citrus reticulata*). Black pit is a disease that affects sweet orange. Both diseases are caused by the bacterium *Pseusomonas syringae* pv. *syringae. P. syringae* pv. *syringae* becomes especially pathogenic for citrus fruits, when the prevailing environmental conditions are high humidity and temperatures around 18°C, which coupled with damage to shoots or fruits by wind, thorns, and hail [34]. Pathognomonic symptoms of disease are water-soaked lesions, which extend from the midrib to the minor ones that surround the base of the petiole. At the last phase of the disease, leaf desiccation and curling are observed. This trait is mainly present in the leaves that remain attached to the stem and finally fall. Necrotic area in twigs expands and finally dies after 4 weeks.

Within that vast number of bacteria that compound the plant microbiome (either rhizosphere or phyllosphere), Pseudomonas bacteria are the most versatile and metabolically varied. As mentioned above, due to its enormous genetic and metabolic plasticity, many species from Pseudomonas genus are successful colonizers of rhizosphere and phyllosphere inducing beneficial effects to the host plants. The most recognized effects are: plant growth promoters, biological control agents, and resistance auto-inducers [35]. Only one species of Pseudomonas is known to be pathogenic for a wide variety of plants; this species is *P. syringae*. P. *syringae* shows a high host plant specificity. Because of this specificity, these strains have been considered as pathovars within the *P. syringae* complex, depending on the type of plant in which the bacterium acts as a pathogen and produces the disease, actually around 50 pathovars are recognized to act as ethological agents in 180 different plant types [36]. *P. syringae* is a leaf-borne commensal bacterium in a wide variety of crop plants, fruit trees, vegetables, and ornamental plants. These bacteria are epiphytic colonizers, which could reach the internal leaf tissue, once in the apoplast bacteria begins its multiplication producing disease symptoms. Therefore, the pathogenesis development is a multi-step procedure such as: (i) entry to internal tissues plant to reaching the intercellular place the apoplast, (ii) to evade plants' resistance responses, and (iii) inducing disease and generating symptoms by particular invasive approaches and molecules [37].

4.1 Quorum sensing in *P. syringae* pv. *syringae*

The quorum sensing system in *P. syringae* uses as auto-inducer a molecule from AHL signaling family: 3-oxo-hexanoyl-homoserine lactone (3-oxo-C6-HSL) molecule. Production of 3-oxo-C6-HSL is dependent on gene *ahlI* that encodes the synthase AhlI. The other component of the quorum sensing circuit is the gene *ahlR*, the signal regulator [38]. When the auto-inducer precursor is available, the synthase AhlI catalyzes the formation of 3-oxo-C6-HSL. Subsequently, the signal regulator AhlR forms a stable complex with 3-oxo-C6-HSL and promotes the transcription of *ahlI* via positive feedback increasing the concentration of 3-oxo-C6-HSL proportionally to cell population density. Additionally, this quorum sensing AhlI/R pathway is subject to effect of regulatory proteins as AefR, this protein actively participates in *ahlI* transcription. A novel regulator, GacA, displays a similar effect in the process of auto-induction. Together AefR and GacA seem to have participated in the activation of the AhlI-AhlR quorum sensing system via independent pathways [38]. The AHL quorum sensing system in *P. syringae* regulates the alginate production, the main component of EPS in *P. syringae. aefR, ahlI, and ahlR* deletion mutant strains display limited survival ability on dry leaves, which is due to the EPS helping epiphytic fitness and desiccation tolerance [39]. Motility is considered an indispensable epiphytic fitness trait, and swarming motility in *P. syringae* is coordinated by a bacterial social behavior [40]. In *P. syringae* swarming motility is regulated by AhlI-AhlR quorum sensing system and AefR. Deletion mutant strains of *aefR* and *ahlI– ahlR–* double mutant display a hypermotile phenotype compared with the wild-type strain [41].

5. Quorum sensing silencing in bacteria: a valuable tool for phytopathogens control

Quorum sensing is a cell-to-cell communication system that depends on population density. This communication system favors the adaptation to new ecological niches, promotes the exploitation of new metabolic resources, and affords competitive advantages to the bacteria that use it. All the aforementioned is directly related to the regulation of virulence, bacterial resistance, and biofilm formation among other phenotypes observed in bacterial population. Quorum sensing disruption is an alternative to reduce the pathogenicity in bacteria. There are two main approaches in order to silence the quorum sensing system in phytopathogenic bacteria: (i) signal degradation/modification termed quorum quenching, (ii) signal overproduction termed pathogen confusion [42]. Quorum quenching is a mechanism approved by several bacteria groups in order to disrupt the QS signaling of contenders, offering to these bacterial cells an additional benefit within a specific niche. It is rational that microorganisms can develop mechanisms to neutralize the QS systems of competing organisms in order to increase their competitive strength in an ecosystem. In a previous study, we have isolated and identified bacterial from citrus phylloplane that display the ability to modify the structure of DSF signal molecule the cis-11-methyl-2-dodecenoic acid in *Xanthomonas citri* subsp. *citri* the etiological agent of bacterial citrus canker. The bacteria *Bacillus vallismortis*, *Pseudomonas oryzihabitans*, *Pseudomonas aeruginosa*, *Raoultella planticola*, *Kosakonia cowanii*, and *Citrobacter freundii* were characterized by molecular technics and display the ability to reduce DSF/rpf communication pathway [43]. We show that these quorum quenching bacteria use a DSF molecule as a substrate for the UDP-sugar transferase enzyme. These bacteria added to DSF molecule one unit of sugar from UDP sugar pools. Thus, the recognition for the RpfC sensor of this modified DSF molecule was impaired. Subsequently, a substantial reduction in the canker lesions was obtained in *Citrus sinesis.*

A previous study used the pathogen confusion approach expressing rpfF from *X. fastidiosa* in sweet orange (*C. sinensis* L. Osb.) using *agrobacterium tumefaciens* in order to reduce citrus canker disease severity. Ectopic expression of xfDSF molecule in *C. sinensis* reduces its susceptibility to *Xcc.* Transgenic plants display a reduction in the number of citrus lesions presumably for the effect on its motility and attachment, also genes involved in the flagella function, pili formation, and T3SS were downregulated in Xcc when they were infiltrated into the leaves of transgenic plants [44].

The bacteria *P. syringae* strain B728a displays the ability to degrade enzymatically different types of AHL. The HacA and HacB are acylases that cleavage the amide bonds of AHL. These enzymes do not have any affect over 3OC6-HSL endogenous accumulation. The heterologous expression of the secreted HacA acylase produced in *P. syringae* strain B728a could become a potential tool in biological control agents, because it might enable the quorum sensing disruption in phytopathogenic bacteria [45].

6. Conclusions

Bacterial coordinated behaviors such as virulence factors production, motility, biofilm formation, and antibiotic resistance are regulated by cell–cell communication system often called quorum sensing. For all aforementioned, quorum sensing silencing has arisen as a good-looking approach to reduce the disease spread and severity. The major bacteria that affect citrus cultivar are endowed with several quorum sensing pathways. The quorum quenching and pathogen confusion implementation approaches could afford new and environmental-friendly strategies for control of this bacterial disease.

Acknowledgements

The authors thank Professor Jesus A Ferro from the Technology Department, Faculdade de Ciencias Agrarias e Veterinarias, Universidade estadual Paulista, UNESP, Jaboticabal, SP, Brasil.

Conflict of interest

The authors declare no conflict of interest.

DOI: http://dx.doi.org/10.5772/intechopen.107902

Author details

Juan Carlos Caicedo* and Sonia Villamizar
Independent Researcher, Colombia

*Address all correspondence to: caicedocepeda@gmail.com

References

[1] Miller MB, Bassler BL. Quorum sensing in bacteria. Annual Review of Microbiology. 2001;**55**:165-199

[2] Federle MJ, Bassler BL. Interspecies communication in bacteria. The Journal of Clinical Investigation. 2003;**112**(9):1291-1299. DOI: 10.1172/JCI20195

[3] Mukherjee S, Bassler BL. Bacterial quorum sensing in complex and dynamically changing environments. Nature Reviews. Microbiology. 2019;**17**:371-382. DOI: 10.1038/s41579-019-0186-5

[4] Waters CM, Bassler BL. Quorum sensing: Cell-to-cell communication in bacteria. Annual Review of Cell and Developmental Biology. 2005;**21**:319-346

[5] Romero M, Mayer C, Muras A, Otero A. Silencing bacterial communication through enzymatic quorum-sensing inhibition. In: Kalia V, editor. Quorum Sensing Vs Quorum Quenching: A Battle with No End in Sight. New Delhi: Springer; 2015. DOI: 10.1007/978-81-322-1982-8_19

[6] Dow JM. Diffusible signal factor-dependent quorum sensing in pathogenic bacteria and its exploitation for disease control. Journal of Applied Microbiology. 2017 Jan;**122**(1):2-11. DOI: 10.1111/jam.13307. Epub November 10, 2016. PMID: 27684652

[7] Lee HA. Further data on the susceptibility of rutaceous plants to citrus canker. Journal of Agricultural Research. 1918;**15**:661-665

[8] daSilva AC, Ferro JA, Reinach FC, Farah CS, Furlan LR, Quaggio RB, et al. Comparison of the genomes of two Xanthomonas pathogens with difering host speciicities. Nature. 2002;**417**:459-463

[9] Moreira LM, deSouza RF, Almeida NF Jr, Setubal JC, Oliveira JC, Furlan LR, et al. Comparative genomics analyses of citrus-associated bacteria. Annual Review of Phytopathology. 2004;**42**:163-184. DOI: 10.1146/annurev.phyto.42.040803.140310

[10] Civerolo EL. Bacterial canker disease of citrus. Journal of Rio Grande Valley Horticultural Society. 1984;**37**:127-145

[11] Yang B, Sugio A, White FF. Os8N3 lateral organ boundaries 1 is a disease susceptibility gene for citrus bacterial canker disease is a host disease susceptibility gene for bacterial blight of rice. Proceedings of the National Academy of Sciences of the United States of America. 2006;**103**:10503-10508

[12] Caicedo JC, Villamizar S. *Xanthomonas citri* ssp. *citri* pathogenicity, a review. In: Khan MS, Khan IA, editors. Citrus—Research, Development and Biotechnology. Rijeka: IntechOpen; 2021. DOI: 10.5772/intechopen.97776

[13] Ryan RP, An SQ, Allan JH, McCarthy Y, Dow JM. The DSF family of cell-cell signals: An expanding class of bacterial virulence regulators. PLoS Pathogens. 2015;**11**:e1004986

[14] Ryan RP, Dow JM. Communication with a growing family: Diffusible signal factor (DSF) signaling in bacteria. Trends in Microbiology. 2011;**19**(3):145-152. DOI: 10.1016/j.tim.2010.12.003

[15] Guo Y, LiYJ ZY, Wang N. Difusible signal factor-mediated quorum sensing

plays a central role in coordinating gene expression of *Xanthomonas citri* subsp. *citri*. Molecular Plant-Microbe Interactions. 2012;**25**:165-179

[16] Barber CE, Tang JL, Feng JX, Pan MQ, Wilson TJG, et al. A novel regulatory system required for pathogenicity of *Xanthomonas campestris* is mediated by a small diffusible signal molecule. Molecular Microbiology. 1997;**24**:555-566

[17] Slater H, Alvarez-Morales A, Barber CE, Daniels MJ, Dow JM. A two-component system involving an HD-GYP domain protein links cell-cell signalling to pathogenicity gene expression in *Xanthomonas campestris*. Molecular Microbiology. 2000;**38**:986-1003

[18] Caicedo JC, Villamizar S, Ferro JA. Quorum sensing, its role in virulence and symptomatology in bacterial citrus canker. In: Gill H, Garg H, editors. Citrus Pathology. Rijeka: IntechOpen; 2017. DOI: 10.5772/66721

[19] Ryan RP et al. Cell–cell signaling in *Xanthomonas campestris* involves an HD-GYP domain protein that functions in cyclic di-GMP turnover. Proceedings of the National Academy of Sciences. United States of America. 2006;**103**:6712-6717

[20] Garcia AL, Torres SCZ, Heredia M, Lopes SA. Citrus responses to *Xylella fastidiosa* infection. Plant Disease. 2012;**96**:1245-1249

[21] Sicard A, Zeilinger AR, Vanhove M, Schartel TE, Beal DJ, Daugherty MP, et al. *Xylella fastidiosa*: Insights into an emerging plant pathogen. Annual Review of Phytopathology. 2018;**56**:181-202

[22] Coletta-Filho HD, Castillo AI, Laranjeira FF, et al. Citrus variegated chlorosis: An overview of 30 years of research and disease management. Tropical Plant Pathology. 2020;**45**:175-191. DOI: 10.1007/s40858-020-00358-5

[23] Nunney L, Hopkins DL, Morano LD, Russell SE, Stouthamer R. Intersubspecific recombination in *Xylella fastidiosa* strains native to the United States: Infection of novel hosts associated with an unsuc- cessful invasion. Applied and Environmental Microbiology. 2014;**80**:1159-1169

[24] Nunney L, Azad H, Stouthamer R. An experimental test of the host-plant range of nonrecombinant strains of north American *Xylella fastidiosa* subsp. multiplex. Phytopathology. 2019;**109**:294-300

[25] Nunney L, Yuan X, Bromley RE, Stouthamer R. Detecting genetic introgression: High levels of intersubspecific recombination found in *Xylella fastidiosa* in Brazil. Applied and Environmental Microbiology. 2012;**78**:4702-4714

[26] Hewitt WB, Houston B, Frazier NW, Freitag JH. Leafhopper transmission of the virus causing Pierce's disease of grape and dwarf of alfalfa. Phytopathology. 1946;**36**:117-128

[27] Severin HHP. Spittle-insect vectors of Pierce's disease virus: II. Life history and virus transmission. Hilgardia. 1950;**19**:357-382

[28] Almeida RPP, Nunney L. How do plant diseases caused by *Xylella fastidiosa* emerge? Plant Disease. 2015;**99**:1457-1467

[29] Baccari C, Lindow SE. Assessment of the process of movement of *Xylella fastidiosa* within susceptible and resistant grape cultivars. Phytopathology. 2011;**101**:77-84

[30] Fouhy Y et al. Diffusible signal factor-dependent cell-cell signaling and virulence in the nosocomial pathogen *Stenotrophomonas maltophilia*. Journal of Bacteriology. 2007;**189**:4964-4968

[31] Newman KL, Almeida RPP, Purcell AH, Lindow SE. Cell-cell signaling controls *Xylella fastidiosa* interactions with both insects and plants. Proceedings of the National Academy of Sciences. 2004;**101**:1737-1742

[32] Chatterjee S, Wistrom C, Lindow SE. A cell-cell signaling sensor is required for virulence and insect transmission of *Xylella fastidiosa*. Proceedings of the National Academy of Sciences. 2008;**105**:2670-2675

[33] Ionescu M, Baccari C, Da Silva AM, Garcia A, Yokota K, Lindow SE. Diffusible signal factor (DSF) synthase RpfF of *Xylella fastidiosa* is a multifunction protein also required for response to DSF. Journal of Bacteriology. 2013;**195**(23):5273-5284. DOI: 10.1128/JB.00713-13

[34] Gorlenko MV. Bacterial Diseases of Plants. 2nd ed. Jerusalem, Israel: Israel Program for Scientific Translations; 1965

[35] Newton AC, Fitt BDL, Ackins SD, Walters DR, Daniell TJ. Pathogenesis, parasitism, and mutualism in the trophic space of microbe-plant interactions. Trends in Microbiology. 2010;**18**:365-373

[36] Gardan L, Shafik H, Belouin S, Broch R, Grimont F, Grimont PA. DNA relatedness among the pathovars of *Pseudomonas syringae* and description of *Pseudomonas tremae* sp. nov. and _*Pseudomonas cannabina* sp. nov. (ex Sutic and Dowson 1959). International Journal of Systematic Bacteriology. 1999;**49**:469-478

[37] Kahlon RS. Pseudomonas-Plant Interactions II: Biology and Pathogenesis of *Pseudomonas syringae*. Springer Nature; 2016

[38] Quiñones B, Pujol CJ, Lindow SE. Regulation of AHL production and its contribution to epiphytic fitness in *Pseudomonas syringae*. Molecular Plant-Microbe Interactions. 2004;**17**:521-531

[39] Yu J, Peñaloza-Vázquez A, Chakrabarty AM, Bender CL. Involvement of the exopolysaccharide alginate in the virulence and epi-phytic fitness of *Pseudomonas syringae* pv. Syringae. Molecular Microbiology. 1999;**33**:712-720

[40] Eberl L, Molin S, Givskov M. Surface motility of *Serratia liquefaciens* MG1. Journal of Bacteriology. 1999;**181**:1703-1712

[41] Quiñones B, Dulla G, Lindow SE. Quorum sensing regulates exopolysaccharide production, motility, and virulence in *Pseudomonas syringae*. Molecular Plant-Microbe Interactions. 2005 Jul;**18**(7):682-693. DOI: 10.1094/MPMI-18-0682 PMID: 16042014

[42] Uroz S, Dessaux Y, Oger P. Quorum sensing and quorum quenching: The yin and yang of bacterial communication. Chembiochem. 2009;**10**(2):205-216

[43] Caicedo JC, Villamizar S, Ferro MIT, Kupper KC, Ferro JA. Bacteria from the citrus phylloplane can disrupt cell–cell signalling in *Xanthomonas citri* and reduce citrus canker disease severity. Plant Pathology. 2016;**65**:782-791

[44] Caserta R, Picchi SC, Takita MA, Tomaz JP, Pereira WE, Machado MA, et al. Expression of *Xylella fastidiosa* RpfF in citrus disrupts signaling in *Xanthomonas citri* subsp. *citri* and thereby its virulence.

Molecular Plant-Microbe Interactions. 2014;**27**(11):1241-1252. DOI: 10.1094/MPMI-03-14-0090-R PMID: 25099341

[45] Shepherd RW, Lindow SE. Two dissimilar N-acyl-homoserine lactone acylases of *Pseudomonas syringae* influence colony and biofilm morphology. Applied and Environmental Microbiology. 2008;**75**:45-53

Chapter 8

Ecology, Biology, Damage, and Management of Sucking and Chewing Insect Pests of Citrus

Muhammad Babar Shahzad Afzal, Ansa Banazeer, Jose Eduardo Serrao, Muhammad Rizwan and Afifa Naeem

Abstract

Citrus are important commodities for human and animal nutrition but these crops are attacked by a plethora of dangerous agents, including viruses, bacteria, fungi, and invertebrates that decrease the yield significantly. Within invertebrates, insects are the more prevalent citrus pests causing plant damage or act as diseases vector. In this chapter, we focused in the insect pests with worldwide distribution in citrus orchards, including sap sucking Asian citrus psyllid *Diaphorina citri*, citrus mealybug *Planococcus citri,* citrus whitefly *Dialeurodes citri*, thrips representatives such as *Scirtothrips citri* and *Pezothrips kellyanus*, and chewing citrus leafminer *Phyllocnistis citrella* and lemon butterfly *Papilio demoleus*. These pests are distributed across various regions of different continents such as in Asia-pacific, Americas, Africa, and Oceana. We presented detailed data from these pests' biology, ecology, damage, and methods for control. The pest incidence and biology is affected by various biotic and abiotic factors thus providing the opportunity to use these factors as method of intervention to disturb pest life cycle. In this context, several IPM techniques such as cultural, physical, biological, and chemical methods were elaborated, which could help to reduce the pest status below damaging levels.

Keywords: citrus, sooty mold, Asian citrus psyllid, citrus greening, citrus mealybug, citrus leafminer, citrus canker, biological control, pest management

1. Introduction

Citrus is one of the largest grown fruit crops in various tropical and subtropical regions of the world, including Brazil, China, USA, India, Pakistan, Italy, Spain, Australia, and Argentina. The genus *Citrus* includes various species of oranges, mandarins, grapefruits, limes, and lemons, which belong to Rutaceae [1, 2]. The production of citrus is limited by biotic (insect pests and pathogens) and abiotic factors (temperature, humidity, soil conditions, and availability of water). Within biotic factors, insects are major constraints in the optimum production of citrus [3–5]. About 250 species have been reported to cause damage to citrus plants but only few pests become a regular problem and cause heavy damage [6]. Overall, the yield losses due to activity of citrus insect pests may reach to 50% without timely diagnosis and management. Citrus insect pests are categorized into sucking and chewing. The eco-biology, damage pattern, and management strategies for each of these citrus pests are discussed in detail in this chapter (summary of citrus insect pests is also given in **Table 1**). Proper pest identification and understanding the eco-biology and damage patterns of citrus insect pests can help to devise and implement suitable pest management program, which can be cost-effective and environmentally friendly and ultimately help to enhance the yield potential of citrus crop.

2. Sucking insects

These insects deprive the citrus plants from essential nutrients by feeding on sap of tender plant parts such as leaves, fruit buds and green stems, and branches using piercing-sucking or rasping-sucking mouthparts. Detail of important sucking insect pests of citrus is given below:

2.1 Asian citrus psyllid

It is also known as citrus psylla. The biological name of this insect is *Diaphorina citri* Kuwayama. It belongs to order Hemiptera and family Liviidae.

2.1.1 Distribution

Diaphorina citri is a serious insect pest of citrus groves in different regions of the world. It was first reported in Taiwan [14] from where it invaded to many citrus-growing regions of the world, including Brazil [15], India [16, 17], Pakistan [18], Southern California [19], Texas [20], Florida [21], Australia [22], Colombia [23], Caribbean [24], Mexico [25], Indonesia [26], Iran [27], Kenya [28], Japan [29], Oman [30], Ethiopia [31], Malaysia [32], and Bangladesh [33].

2.1.2 Eco-biology

Diaphorina citri is mainly a pest of tropical and subtropical climate. New foliage growth (flush) regulates the dynamics of *D. citri* requiring soft tissues for oviposition and development [34]. The peak period of psyllid is coinciding with new flush and pest becomes active at the end of February (late winter) and population reaches at maximum level in March and April (spring season). The adults become active in May (early summer) and June and new colonies develop in July (mid-summer). The population begins to decline in October (autumn) and only fraction of population is observed at the end of December and January (winter season) [35]. Reproduction of *D. citri* is totally dependent on availability of young shoots containing feather stage to recently expanded tender leaves. Adult females feed on tender shoots to mature their eggs and prefer opening buds and emerging shoots for oviposition. During the following 2–3 weeks, shoot and leaf tissues are still tender and are utilized by nymphs and adults to complete development and mature eggs, respectively [34, 36]. Adults can also feed and survive on the fully developed leaves for several months. Temperatures between 24°C and 30°C are most favorable for both adult survival and reproduction, as adults survive for 30–50 days and females lay about 500–800 eggs at these temperatures [37, 38]. The damage caused by *D. citri* is most severe in autumn than in

Common name	Species name	Family	Mode of feeding	ETL*
Asian citrus Psyllid	*Diaphorina citri*	Liviidae	Piercing-sucking	(i) 0.5–1.0 adult psyllid per stem tap [7] or (ii) 6 nymphs or adults per leaf [8]
Whitefly	*Dialeurodes citri*	Aleyrodidae	Piercing-sucking	20–30 nymphs per leaf on oranges and lemons and 5–10 nymphs per leaf on mandarin species [9]
Mealybug	*Planococcus citri*	Pseudococcidae	Piercing-sucking	(i) 5–10% fruit infestation by colonies of young nymphs in summer and (ii) 15% fruit infestation in autumn [10]
Thrips spps.	*Scirtothrips citri* *S. aurantii* *S. dorsalis* *S. inermis* *Pezothrips kellyanus* *Megalurothrips kellyanus* *Thrips major*	Thripidae	Rasping-sucking	10 thrips/branch tapping [11]
Red scale	*Aonidiella aurantii*	Diaspididae	Piercing-sucking	3–5 nymphs per leaf [12]
Blackfly	*Aleurocanthus woglumi*	Aleyrodidae	Piercing-sucking	5–10 nymphs per leaf [13]
Leafminer	*Phyllocnistis citrella*	Gracillariidae	Larvae burrows beneath the leaf epidermis and consume the leaf tissues by feeding in serpentine manner	10% infestation [8]
Lemon butterfly	*Papilio demoleus*	Papilionidae	Larvae feed upon both young and mature leaves and cause defoliation	3–5 larvae per plant [13]
Fruit fly species	*Bactrocera zonata* *Ceratiitis capitata* *B. tryoni*	Tephritidae	Larvae burrow inside the fruit and consume the fruit pulp	20 adults per trap per week [12]

**Economic Threshold Level.*

Table 1.
Summary of citrus insect pests.

summer. The order of severity varies due to type of host plant, age, abundance of flush, variation in flush phenology, and management practices [39].

D. citri has three developmental stages of life, for example, egg, nymph, and adult. Bindra [40] observed almond-shaped eggs that are yellow in color and are laid during day time either isolated or in masses of double or triple lines embedded into leaf tissue with short stalk. *D. citri* has five nymphal instars that are orange to yellow in color, with rounded flattened body (**Figure 1**) [35]. Pande [41] studied the biology of *D. citri*, finding that mating onset just after adult emergence.

The egg laying capacity per female is about 180–520. The incubation period ranges from 4 to 18 days according to environmental conditions. Nymphs have five instars, which take 10–30 days to adult molt. The adults survive longer and take 14–48 days to complete their life cycle. *D. citri* has 10 overlapping generations per year [41].

2.1.3 Damage

Diaphorina citri is one of the devastating sucking insect pests of citrus plantations and losses due to attack of this pest ranges from 82 to 95% [42]. Both the nymphal and adult stages of this pest feed on juvenile plant shoots and leaves and floral buds and deprive the plant from essential nutrients by sucking phloem sap. Its infestation results in leaf curling, distortion, and shedding of flower and leaves [43, 44]. They also inject toxic saliva in the citrus plants during feeding [43]. The host range of *D. citri* has been known to be *Citrus* spp. and near relatives, spanning over 23 genera within the Rutaceae [45]. During its feeding, *D. citri* (4th and 5th instars and adults) transmits gram negative and phloem-limited bacterial pathogens (*Candidatus* Liberibacter *asiaticus, Candidatus* Liberibacter *americanus,* and *Candidatus* Liberibacter *africanus*) to citrus plants. These pathogens are the causal agents of devastating disease in the citrus plantations, which is known as citrus greening or Huanglongbing (HLB) or yellow dragon disease [46, 47]. The citrus greening is one of world's most serious diseases of all citrus cultivars. This disease is a threat for citrus industry throughout the world as it has perished millions of hectare of citrus in about 40 countries [48]. This disease results in poor quality fruit production, severe yield

Figure 1.
Scaly eggs of Diaphorina citri *laid singly in a row below the leaf surface (A), nymphal instars of* D. citri *congregated for feeding at soft branch (B), adult* D. citri *with elaborated body features and sitting in upright position with rear raised (C), and adults of* D. citri *infesting the underside of leaf surface and secreting honeydew and wax (D). Photographs by Douglas L. Caldwell, University of Florida, USA (A and C), https://www.koppert.com/challenges/pest-control/psyllids/asian-citrus-psyllid/ (B) and the citrus pest and disease prevention program, California Department of Food and Agriculture, USA (D).*

reduction (30–100%), and ultimately death of whole plant within 5–8 years of attack [49–51]. Other symptoms of HLB infection are progressive blotchy mottling of leaves, plant stunting, off-season bloom, deformed, small-sized, and off-flavor fruit with high acid contents and bitter taste and premature fruit drop thus inducing market losses in fresh and processed fruits [52–54]. Asymmetrical patterns with yellow veins on leaves are also clear indication of citrus greening. In addition to these, several other symptoms such as stunted plant growth, shoot dieback, fruit drop, and overall yellow appearance of citrus plants are visible signs of HLB infection [55]. Before the appearance of visible symptoms, the initiation of root dieback and decrease in root-shoot ratio have also been observed in HLB-infected plants [56].

2.1.4 Management

2.1.4.1 Biological control

Biological control is the control of insect pests using predators and parasitoids and should be adapted at large scale to avoid the unnecessary use of insecticides. There are various predators and parasitoids of *D. citri* in the citrus crop. The primarily source of generalist predators are syrphid flies, ladybird beetles, lacewings, predatory mites, and ants [57]. The immature of *D. citri* are attacked by ants [58]. Husain and Nath [35] and Batra et al. [59] also reported that different species of coccinellid attack the *D. citri* such as the seven-spot ladybird beetle *Coccinella septempunctata* Linnaeus, transverse ladybird beetle *C. repanda* Thunberg, Malaysian ladybird beetle *Chilocorus nigrita* (Fabricius), zigzag ladybird beetle *Cheilomenes sexmaculata* (Fabricius), and three-striped lady-beetle *Brumus suturalis* Fabricius. The larvae of syrphid fly *Allograpta* spp. have been reported to attack *D. citri* nymphs in the regions of Nepal and Reunion [60]. The primary effective parasitoids of *D. citri* are *Tamarixia radiata* (Waterston) (Hymenoptera: Eulophidae) [61] and *Diaphorencyrtus aligarhensis* (Shafee, Alam and Argarwal) (Hymenoptera: Encyrtidae) [62], which are native to India and provide better control than predators. Female of *T. radiata* prefers the 3rd, 4th, and 5th nymphal instars of *D. citri* [63, 64].

2.1.4.2 Chemical control

There is significant increase in insecticide use per year to control *D. citri* and citrus greening and annual cost for managing this pest could range from $US 240 to>$US 1000 depending upon application frequency, type of insecticide sprayed, and method of application [65]. Among various types of insecticides, the foliar applied broad-spectrum insecticides are recommended to control the *D. citri* prior to flushing (to kill the overwintering adults) and during growing season. Foliar sprays with broad-spectrum insecticides such as chlorpyrifos, dimethoate, fenpropathrin, bifenthrin, and zeta-cypermethrin along with foliar and soil applications with systemic neonicotinoid insecticides such as imidacloprid, thiamethoxam, and clothianidin are indicated to control *D. citri* [66–68]. Imidacloprid offers 50–90% control of adult psyllid population in the field (reviewed by Boina and Bloomquist, [48]). Spray of imidacloprid in rotation with chlorpyrifos or cypermethrin at two-week interval also reduces the psyllid population and incidence and spread of HLB during new flush stage of citrus plants [69]. In the world, all citrus-growing areas are not free from attack of citrus greening and its vector. There is no permanent cure for controlling this pest except chemical control, which keeps the pest population at low level and is one of most effective management options for controlling the pest incidence and spread of citrus greening disease.

2.1.4.3 Miscellaneous tactics

Chemical and biological control should be combined with other control measures such as promoting a clean cultivation, which includes using a disease free and transgenic plants that are resistant to *D. citri* and removal of infected plants from the field; and use of antibiotics (e.g., tetracycline hydrochloride) for suppression of citrus greening symptoms is viable and sustainable tool to control the incidence of HLB [70]. To reduce the insecticides resistance in the *D. citri*, there is an immediate need for development of IRM (insecticide resistance management) strategies. Regular monitoring of *D. citri* in field to determine the rate of development of insecticide resistance is one of the major components of IRM. In addition, resistant colony should be developed in the laboratory for determining the genetic nature, mode of inheritance, stability of resistance, fitness costs, and pattern of cross resistance in order to manage insecticide resistance.

2.2 Citrus whitefly

Citrus whitefly (CWF) belonging to order Hemiptera and family Aleyrodidae is technically known as *Dialeurodes citri* (Ashmead).

2.2.1 Distribution

Dialeurodes citri has a wide range of distribution in different regions around the world [71]. With its origin in Southeast Asia, CWF occurrence was reported from southeastern United States mainly in Florida in the 1880s [72] from where in 1900 it spread to California, a region in the western United States [73]. Later its occurrence was recorded in the Mediterranean countries such as Western Galilee region of Israel in 1975 [74, 75] and in the Turkey's Eastern Mediterranean citrus groves in 1976 [76, 77]. In 1977, invasion of this species in citrus orchards was reported from South Adriatic, near Dubrovnik [78]. This species has also been detected from Oceania, New Zealand, in 2000 [79]. In Asia, this species is infesting citrus orchards in several parts of Pakistan [80], Taiwan, Japan, and China [81, 82], Uzbekistan, and Turkmenistan [83], and India [84, 85].

2.2.2 Eco-biology

The eggs of CWF are yellow and have smooth surface. Female adults lay their eggs on leaves and 8–24 days are required for hatching according to climate [86]. The CWF is an arrhenotokous species [87] in which unfertilized eggs always develop into males. The nymphs are elliptical, flat, and scale like. After first molt, the instars become fixed (legs and antennae are lacking) at the underside of the leaf surface until the adult stage. About 23–30 days are required to complete the nymphal period. The pupa of CWF is opaque and eye spots of developing adult are clearly visible from pupal integument and pupa completes its development in about 13–30 days. The total life cycle from egg to adult formation is completed in 41–333 days according to the ecological conditions such as temperature, humidity, and rainfall. The CWF overwinters as nymph (fourth instar) at the underside of the leaves. The pupae appear early in spring and in March-April (late spring) the adult emergence occurs [86]. The number of generations varies according to the region with 2–5 per year [75, 88].

2.2.3 Damage

The CWF has sucking mouth parts and both nymphs and adults injure the plants by sucking sap. The further injury is caused by honeydew release, which results in the development of sooty mold fungus over fruits and foliage [86]. The sooty mold may cover the leaves and cause indirect damage by interfering with respiration and photosynthetic activity of plants, which leads to leaf drop and yield reduction [75]. The infested citrus trees become weak and tasteless. Easy peeler citrus varieties such as mandarins and sweet orange cultivars are preferred citrus hosts of CWF [87]. Besides this, CWF has also been reported as a vector of disease known as Citrus Yellow Vein Clearing Virus (CYVCV) [89]. This viral disease was reported for the first time from Pakistan in 1988 in sour orange (*Citrus aurantium* L.) and lemon (*C. limon* Burm.f.) [90]. Now this disease is widely distributed in major Chinese province of citrus-growing areas and considered to be most serious disease, which affects the lemon production [91]. The major symptoms of CYVCV that appear in sour orange and lemon are severe vein clearing, vein necrosis, and leaf distortion [91–93]. Usually, this virus does not cause tree death but it can reduce the yield for example in Anyue, Sichuan Province of China up to 80% lemon production is affected by CYVCV and yield is reduced by 50–80% [89].

2.2.4 Management

2.2.4.1 Biological control

The biological control of CWF is poorly studied but it includes mainly the use of potential predatory insects. The main natural enemies of CWF include representatives of predatory coccinellid with highlight to *Serangium japonicum* Chapin (Coleoptera: Coccinellidae). This natural enemy has been reported in pesticide-free crops with many CWF, in the Japan with peaks in May and July reducing the CWF number in the second peak [94]. Release of another ladybird beetle *S. parcesotum* Sicard (Coleoptera: Coccinellidae) in the East Mediterranean region reveals that the predator has success in the colonization of orchards with potential to control CWF [77, 95]. Some parasitoids also have potential to control CWF, including *Encarsia lahorensis* (Howard) (Hymenoptera: Aphelinidae), and *Eretmocerus debachi* Rose and Rosen (Hymenoptera: Aphelinidae) [96–98] but further studies to optimize the control are necessary. Likely parasitoids, pathogenic agents are also poorly evaluated. The fungus *Aschersonia placenta* Berkeley and Broom (Hypocreales: Clavicipitaceae) isolates were evaluated in China and three of them have potential to control CWF [82]. In addition, *A. aleyrodis* Webber (Hypocreales: Clavicipitaceae) was reported causing high mortality in *D. citri* at Southern Alabama and China [99]. *Lecanicillium attenuatum* Zare and Gams from order Hypocreales have also been reported to kill the nymphs of CWF, thus considered a potential biological control agent of this pest [100, 101].

2.2.4.2 Physical control

Physical control involving the use of ultraviolet (UV) light to control insects mainly CWF is gaining importance as one of the components of IPM because this technique is environmentally benign and non-hazardous to non-target organisms [102–105]. Traps designed with UV releasing tubes can be installed in the field to monitor and reduce the population of CWF as adults of this pest exhibit positive phototactic behavior to UV source [102, 106, 107]. UV light kills the captured insects by inducing oxidative stress and altering some life traits such as behavior, developmental patterns, and biochemistry [108–110]. Exposure of CWF for longer period, for example, about seven hours per day, can decrease the fecundity, and oviposition rate. Moreover, pupal formation, longevity of adult females, and adult emergence have also been reported to decrease significantly in CWF upon exposure to UV light [107]. Apart from UV light, colored sticky cards and yellow sticky traps should also be used during the active season of CWF in orchards to detect, monitor, and control the insect population as a part of IPM [111, 112].

2.2.4.3 Cultural control

All the practices in citrus orchard that enhance the passage of air flow through the canopy of citrus trees come under the cultural control, these practices include the following: maintaining a proper plant to plant and row to row distance, weed eradication, light to moderate pruning, and optimum application of irrigation and fertilizer. These cultural practices do not allow humidity among the trees to increase significantly and thus keep the population under check [113, 114].

2.2.4.4 Chemical control

Chemical control using inorganic compounds, botanicals, and synthetic insecticides is an integral part of IPM for the control of CWF; however, it should be used judiciously and only when required. Two applications of summer oil or white oil (petroleum) emulsion are recommended during peak activity of CWF; however, in case of very high populations density spray can be done 3–4 times, Refs. [113–115] suggested that use of pyriproxyfen or buprofezin @0.05% twice with the span of 45 days can give good control of CWF in citrus orchards. The joint application of an organophosphate insecticide triazophos and neem formulation can suppress the nymphs and adults of CWF if sprayed twice at 15-day interval in the citrus orchards [116]. Use of tree spray oil (0.5 and 1%) along with lime sulfur diluted with water can provide satisfactory control of CWF eggs mainly in the orchards of sweet oranges. Moreover, significant nymphal control of CWF can be achieved by spraying 0.03% dimethoate and formothion [117]. Sole reliance upon conventional chemical control of CWF should be avoided as this pest can develop resistance quickly to different pesticides render them ineffective [118, 119].

2.3 Citrus mealybug

The scientific name of citrus mealybug (CMB) is *Planococcus citri* (Risso) and this species belongs to order Hemiptera and family Pseudococcidae.

2.3.1 Distribution

CMB is polyphagous and most destructive pest of citrus orchards and nurseries. It is distributed in different parts of the world such as in Egypt [120], Florida [121], California [122], Portugal [123], Turkey [124], South Africa [125], USA [126], South pacific region [127], Australia [128, 129], India [130], Mediterranean region [131, 132], America [122, 133], and Pakistan [134].

2.3.2 Eco-biology

Sexual dimorphism occurs in CMB [135]. The female of CMB has oval, flat, and soft body, which is covered with wax and long waxy filaments [136]. The length of adult female insect is about 3 mm [135]. Similar to other scale insects, females of CMB are wingless [137]. Female lays yellow eggs within ovisacs in soil and each ovisac contains about 300–800 eggs. About 10–20 days are required to egg hatching. After hatching, the amber colored nymphs emerge and start feeding by inserting the mouthparts in the lower side of leaves epidermis. The female molts three time to become an adult and complete its nymphal duration in six to eight weeks. The male is gray, winged, and midge like and has long antennae but no mouthparts (**Figure 2**) [138]. Adult males are 1 mm in length and have two caudal filaments [135]. In its early stage, male resembles with female but as it grows, male secretes fibrous, cottony cocoon from which adult male emerges [139]. The male nymphal instars molt 4 times until adult [138]. During winter, they hide themselves in cracks and cavities in tree trunks in the adult female or egg stage. In late spring according to the temperature, especially at the end of April or beginning of May (early summer), they emerge from hibernating sites. CMB has about three to six overlapping generations per year but the Spring-Summer life cycle is major concern for citrus growers as peak infestation occurs in the month of June in Mediterranean region of the Turkey [140]. The males of CMB exhibit the phenomena of polygyny and can fertilize multiple females during their short life span [141, 142]. Generally, the sperm from younger males has more chances to fertilize the female eggs [143].

2.3.3 Damage

The CMB has potential to cause high loss in agriculture especially in citrus and grapevine industry. The CMB infestation causes direct and indirect type of damage to citrus orchards [144]. All parts of tree such as new leaves, stem, flowers, and fruits are damaged by CMB except underground portion of the plant. The CMB (nymphs and adult female) sucks the sap from different parts of plant, which results in defoliation (about 80%), wilting, dropping of fruits and flowers, deformed fruit appearance [132], and premature yellowing and causes approximately 95% loss in crop yield [145, 146]. In case of heavy infestation, stunting and death of plants occur [136]. According to Smith et al. [128], in South Africa, the early ripening cultivars are more susceptible to damage by CMB than the late maturing cultivars. In early ripening cultivars, the natural enemies have less opportunity to suppress the pest population of CMB in the field before harvesting stage. Generally, fruit damage is caused by CMB between petal fall and at that time when fruits are size of golf ball [147]. The indirect damage to plant by CMB is the production of copious amount of honeydew on foliage and fruits, which provides the growth medium for development of sooty mold fungus [145]. Sooty mold is dark superficial coating on different parts of plants, which decreases the photosynthetic capacity of plant by reducing the amount of light entering the leaf cells; the presence of this black mass upon harvested fruit ultimately reduces the market value [144]. Honeydew is also source of food for ants, which may protect the CMB from their natural enemies [148].

Figure 2.
Planococcus citri *adult male having membranous wings and tail filaments (A), cottony ovisac with egg mass deposited by adult female of* P. citri *(B), nymphs, pre-pupae and pupae of* P. citri *(C) and adult female of* P. citri *whose body is covered with white powdery mass (D). Photographs by lance S. Osborne, the University of Florida, USA (A), Lyle J. buss, University of Florida, USA (B) Paul J. Johnson, South Dakota State University, USA (C) and William Bodine in 1455 N Val vista Dr., Mesa, AZ 85213, USA captured on august 19 2022 (D).*

2.3.4 Management

2.3.4.1 Mechanical and cultural control

During winter, pruning of infested parts is recommended as it allows sufficient light penetration in the canopy, and thus helps in exposing the hibernating insects which are ultimately killed by natural enemies and solar light. Pruning as prophylactic approach is also best management option to prevent the attack of CMB in next season [10, 149]. Some ant species in the citrus orchards hinder the success of biological control of CMB by disrupting the activity of predators and parasitoids and they also act as mechanical carrier for mealybugs and transport them to their feeding sites [148]. Therefore, destruction of the ant colonies in citrus orchards is suggested to disrupt the mutualism between ants and CMB [10, 128, 137]. Plowing the soil near tree trunk during summer to expose the eggs and females to their natural enemies and sunlight is also best technique to control this pest. The sticky bands of 7–8 cm should be wrapped around trunk during second week of December at the height of 0.5 m from ground. The population of CMB can also be controlled by removing the bark as it helps in the elimination of harboring sites of CMB nymphs [150].

2.3.4.2 Use of host plant resistance

The tolerance/resistance against insect pest is also important phenomena and has been observed in three citrus cultivars *viz.*, *Citrus limon*, *C. macroptera,* and *C. grandis* against the attack of CMB. Therefore, these cultivars should be used as rootstock in management program against CMB. *C. sinensis* and *C. limettiodes* have been found moderately resistant to the attack by CMB and only suffer 20.65–30% leaf infestation [151].

2.3.4.3 Biological control

2.3.4.3.1 Predators

The mealybug destroyer *Cryptolaemus montrouzieri* Mulsant (Coleoptera: Coccinellidae) from Australia was introduced into California in 1982, which effectively controlled the CMB infestation in field. This coccinellid predator provides the effective control by reducing the eggs masses, nymphs, and adult population of CMB [152, 153] but Hattingh and Tate [154] found that *C. montrouzieri* is sensitive to IGR, pyriproxyfen, which is used for control of red scales in citrus orchards. It is recommended to release about 500 *C. montrouzieri* beetles per acre in the citrus orchard for effective control [155]. The other most common predators of CMB include *Scymnus syriacus* (Mars.) (Coleoptera: Coccinellidae) and the chrysopid predator, *Chrysoperla carnea* (Stephens) (Neuroptera: Chrysopidae) [156]. Efforts should be made to conserve and enhance the effectiveness of these predators in the orchards by relying less on chemical control.

2.3.4.3.2 Parasitoids

The hymenopteran wasps are most abundant biological control agent for CMB in various regions of the world. The two parasitoids *Leptomastix dactylopii* Howard and *Leptomastidea abnormis* (Girault) (Hymenoptera: Encyrtidae) are used as bio-control insects against CMB [157–159]. However, Mani [160] reported that the indigenous, *Coccidoxenoides peregrines* (Timberlank) (Hymenoptera: Encyrtidae) is more abundant and having parasitism rate of about 10 to 30%. Krishnamoorthy and Singh [161] reported the successful example of release of *L. dactylopii* in India in 1983 on mandarins. Within two month of release, it provided 100% parasitism rate by feeding on all stages of CMB. In South Africa, *C. peregrines* are reared at commercial level and provide good control against CMB [147].

2.3.4.3.3 Pheromone

A synthetic pheromone [(1R-cis)- 2, 2-dimethyl −3-(1-methylethenyl) cyclobutyl methyl acetate] of CMB elicits a positive response against male adults and half-life of this pheromone is about 2 weeks in field and maximum males are caught at the dose of 400–700 mg [162].

2.3.4.3.4 Entomopathogenic nematodes

The entomopathogenic nematodes (EPN) have ability to control the wide range of insect pests including CMB and can be applied as bio-control agent. Once they penetrate inside the host haemocoel and release symbiotic bacteria, the host is killed within 24–48 h. EPN have no negative effect on environment, human, and other vertebrates. The six indigenous species have been used to access the susceptibility of CMB. Three heterorhabditids species (*H. zealandica*, *H. safricana*, and *H. bacteriophora*) and three steinernematids (*S. yirgalemense*, *S. khoisanae*, and *S. citrae*) were evaluated. *The S. yirgalemense* and *H. zealandica* caused highest mortality of about 97 and 91%, respectively, as compared to the other four tested species of nematodes [125], therefore these two species should be incorporated in the biological control program of CMB.

2.3.4.4 Chemical control

The management strategies for control of CMB have been mainly based upon bio-control agents. However, the control with chemical method is most common and widely used strategy because of poor adaptation of natural enemies in varying climatic conditions [10]. The control of CMB by using chemicals can be difficult and effectiveness of chemicals depends on correct application at time when needed. However, acceptable control of CMB may not be achieved with single treatment; the follow-up application of chemicals is necessary. The effective control can be achieved if application is started at the time of initial infestation of pest population [163]. To preserve the natural enemies, it is mandatory to use selective insecticides for control of CMB [164]. Kütük et al. [124] conducted a study to evaluate the effect of biological and chemical control against CMB. They found that summer oil and spirotetramat are compatible with natural enemies, for example, *C. montrouzieri* and *L. dactylopii*, while these natural enemies showed incompatibility with chlorpyrifos-ethyl, due to its side effect on these bio-control agents. Due to cryptic nature and protection with waxy material of CMB, it is necessary to use the chemicals with high vapor pressure. The major insecticides recommended for the control of CMB are from organophosphate (e.g., chlorpyrifos, malathion, dimethoate, azinfos-methyl, dichlorvos parathion, diazinon, and pirimiphos-methyl) and carbamate group (e.g., methomyl, thiodicarb, and carbaryl). These insecticides are applied individually or are mixed with mineral oils [141, 165]. The excessive use of these insecticides for the control of CMB might lead to development of insecticide resistance; therefore, some alternative insecticides such as dinotefuran, acetamiprid, imidacloprid, and thiamethoxam from neonicotinoid group can be employed in the management program of CMB [141].

2.4 Citrus thrips

Citrus thrips is piercing-sucking insect, which belongs to order Thysanoptera and family Thripidae. Various species of phytophagous thrips have been reported upon citrus in the world such as *Scirtothrips citri* (Moulton), *S. aurantii* Faure, *S. dorsalis* (Hood), *S. inermis* Priesner, *Pezothrips kellyanus* Bagnall, *Megalurothrips kellyanus* (Bagnall), western flower thrips *Frankliniella occidentalis* (Pergande), *F. bispinosa* (Morgan), *Thrips major* Uzel, *T. hawaiiensins* Morgan, *T. meridionalis* (Priesner), *T. angusticeps* Uzel, *T. obscuratus* (Crawford), *T. tabaci* Lindeman, *Heliothrips haemorrhoidalis* Bouché, *Chaetanaphothrips signipennis* (Bagnall), and *C. orchidii* Moulton [166–175].

2.4.1 Distribution

Different species of thrips reported on citrus have worldwide distribution. Among all the species, *S. citri* commonly known as citrus thrips or California citrus thrips is of great economic importance in USA and Asia. In USA, it is infesting citrus particularly oranges (Navel oranges), grapefruit, and lemons in California, Arizona, northwestern Mexico, Florida, Washington, and Georgia [174, 176–179]. In Asia, *S. citri* has been recorded in India, China, Iran, and Thailand [179, 180]. Another thrips of economic importance on all type of citrus plantations (lemons, oranges, and grapefruits) is *P. kellyanus*, commonly known as Kelly's citrus thrips (KCT). *P. kellyanus* is Mediterranean in its origin and has been distributed in various geographical regions around the world in Australia [170, 181], New Zealand [182], Southern France [183], Cyprus [171], Italy [184, 185], Spain [186], Turkey, and Greece [187, 188], Portugal [189], Chile [190], Morocco [191], and Tunisia [175].

2.4.2 Eco-biology

Citrus thrips complete their life cycle from egg to adult stage in two to three weeks. The duration of each stage varies with the host species attacked, as well as temperature and humidity of the environment [192]. Citrus thrips is hemimetabolous insect and has following stages: eggs, 1st and 2nd instars or larvae (active feeding immature stages), 1st and 2nd stage pupae (non-feeding instars known as pre-pupa and pupa), and adult or imago (equipped with fringed wings) [176]. The adult female prefers to lay eggs deeply inside the soft and young tissues of leaves, stems, and floral buds with the help of saw-like ovipositor; the egg laying-puncture is completely closed after egg laying. On an average 25–35 eggs are laid per female with maximum of 250 during its life time. Fertilized eggs mostly develop into females and seldom to males. However, only male offspring is developed from unfertilized adult females. Eggs hatch in 5–8 days depending upon the temperature. First and second instars are active feeders and complete their development in 4–14 days by feeding upon soft leaves and under the sepals of tiny fruits. The third and fourth instars known as pre-pupa and pupa, respectively, are passive stages and do not feed, and they spend their time on the ground, tree crevices or inside the curled leaves. Development period from egg to adult formation varies according to temperature such as it requires 16 and 13 days at 25°C and 31°C, respectively. The adult female lives for 26–30 days at 31°C. The number of generations of citrus thrips ranges from 8 to 12 in various agro-climatic conditions. The activity of thrips commence usually in spring and at the beginning of summer at the temperature range of 20–25°C as it coincides with the flowering period and formation of young fruits on trees. The rise in temperature above 30°C and absence of flowers and young fruits result in decrease in thrips population density [175, 177, 193–196].

2.4.3 Damage

Citrus thrips attack usually commence during the flowering stage mostly at petal fall or new fruit formation. Thrips nymphs (1st and 2nd instars) and adults with the help of their asymmetrical piercing-sucking stylet inflict damage to citrus leaves, flowers, buds, and young fruits by extracting the cell contents. Feeding activity upon newly developed citrus flush often results in curling of leaves, along with the appearance of silvery or grayish scars upon the leaves. The damage also results in poor development and growth of infested plant [197–199]. Fruit scarring is the diagnostic damage pattern of citrus thrips, which usually starts when fruit is developing (≤2.5 cm diameter) (**Figure 3**).

Due to thigmotactic nature of citrus nymphs, they prefer to feed under the calyx of young fruit, which results in the development of ring-shaped scar around the calyx or fruit peduncle. In some citrus cultivars such as Navel oranges, thrips activity develops the scaring marks at the styler end of the fruit surface. "Russetymarking" and "tear staining" upon the peel or ring of citrus fruits such as in oranges is also sign of thrips induced damage. Scarring marks are mostly observed upon fruits, which lie outside the canopy. Thrips herbivory upon developing citrus fruits also causes abortion, irregular development or deformity, discoloration, and necrosis of the fruit [200]. Blemishes and scar patterns (cosmetic losses) due to feeding of citrus thrips enormously reduce the market value of citrus fruits leading to export and monetary losses to citrus growers and exporters [184, 201, 202]. Abrasion and scarring of citrus fruits by citrus thrips negatively affects the physiochemical qualities of fruits such as higher percentage of TSS and sugar/acid ratios, lower titratable acidity, increase in water loss, and more rapid weight loss [203].

2.4.4 Management

2.4.4.1 Biological control

The use of predatory mites as biological control agent of citrus thrips is well documented. The most important predacious mites recommended against citrus thrips are phytoseiid mites: *Euseius tularensis* Congdon and *E. hibisci* (Chant) (Acari: Phytoseiidae) and anystid mite: *Anystis agilis* Banks (Acari: Anystidae). These species can be conserved and released in mature citrus orchards and scarring of citrus fruits can be minimized [204–209]. Some other predacious mite species such as *Iphesius* (= *Amblyseius*) *degenerans* Berlese, *Neoseiulus barkeri* (Hughes) (Hughes) (= *Amblyseius mckenziei*), and *N. cucumeris* (Acari: Phytoseiidae) can be used as potential biological control agents against immature thrips in citrus nursery [210]. Another predatory mite *Gaeolaelaps aculeifer* (Acari: Laelapidae) has been shown to successfully reduce the infestation of thrips in commercial citrus orchards of Valencia, eastern Spain after the augmentative release [211].

Figure 3.
Citrus fruit (Kinnow mandarin) infested by thrips showing scarring mark as one of the diagnostic damage patterns. Photograph by Bodil N. Cass, Department of Entomology and Nematology, University of California, Davis, USA.

The use of natural insect predators is advocated for the control of thrips in citrus orchards in different regions of the world. Among the generalist predators of thrips, *Orius insidiosus* (Say) (Hemiptera: Anthocoridae) commonly known as minute pirate bug is an aggressive and voracious predator of thrips adults and nymphs in citrus [173, 179, 212, 213]. Some other generalist thrips predators known are chrysopids, coccinellids, and mirids [214]. Two predatory thrips species *Karnyothrips flavipes* (Jones) and *Leptothrips* spp. have been identified and are known to attack thrips in citrus in Florida, USA [215].

2.4.4.2 Chemical control

Complete thrips control in citrus orchards cannot be obtained merely by biological control agents; therefore, application of different insecticides for the suppression of thrips population below economic damaging levels is unavoidable. The insecticides from organophosphorus (OP), pyrethroids, and new chemical groups are recommended for thrips control. Spray with systemic insecticides at pre-flowering or post-flowering stage is recommended. Dimethoate, a systemic and broad-spectrum OP insecticide, is highly recommended and widely used insecticide against thrips in citrus orchards. It is applied before petal fall when less than 10% buds have opened. Its application on nursery plants is not recommended; however, mature fruits can be sprayed only twice with this insecticide. Among the pyrethroid group insecticides, one application of cyfluthrin or fenpropathrin (broad spectrum) in a year is recommended in the citrus orchards to trees ≤3 years of age. Application of two different pyrethroid insecticides in sequence must be avoided in order to minimize the resistance development. OP and pyrethroid insecticides should be used in rotation as a part of insecticide resistance management tactic. In California, chemical control is initiated when 75 percent petal-fall is complete. In order to minimize the impact of insecticides on natural enemies, and to reduce the resistance development against OP and pyrethroid insecticides, some botanical or microbial nature insecticides such as sabadilla, abamectin, and spinosad should be used for spray. Among the new chemistry insecticides, imidacloprid, acetamiprid, and thiamethoxam from neonicotinoid group and chlorfenapyr from pyrroles group are also recommended against citrus thrips [177, 216–219].

3. Chewing insects

3.1 Citrus leafminer

The scientific name of citrus leafminer (CLM) is *Phyllocnistis citrella* Stainton and it belongs to order Lepidoptera and family Gracillariidae.

3.1.1 Distribution

The CLM is a most destructive pest of citrus nurseries, top grafting trees, and also citrus growing in plastic greenhouses in various parts of the world [220]. This pest is native to eastern and southern Asia occurring in China, India, Thailand, Japan, and Vietnam [221, 222]. It was reported in southern Florida in 1993 [223, 224], most of Caribbean region and Southern part of USA [225, 226]. It has also been reported in Northern part of South America, Uruguay, and Mediterranean coast of Europe [227], in the Middle East [228], Reunion Island [229], and North Africa [227]. In 1999, its damage was reported from Arizona [230].

3.1.2 Eco-biology

The number of eggs laid per female is about 30–75. The eggs are translucent, white, and appear similar to tiny water droplets. Usually female lays eggs underside of leaves and along the leaf mid-vein toward leaf petiole. Eggs hatch in 2–10 days according to the environmental conditions. CLM has four larval stages and having total developmental time of about 5 to 20 days. Upon hatching, larvae immediately begin mining beneath the epidermal cells. The 1st instar larvae are green in color, translucent, and difficult to detect. The 2nd and 3rd instar larvae are also translucent and yellow-green in color. The 4th instar larvae are clearly visible and form silken cocoon within mines. 4th instar larva curls the leaf edge as silk dries over the cocoon and forming a protective shell called pupa. In initial stage, pupa is yellow-brown in color but at later stage it becomes darker. About 6–22 days are required to complete the pupal duration. The adult is white and has silvery scales on the dorsal surface of forewings and distinctive black spots on the tip (**Figure 4**). The moth is small and of the size of mosquito and active during early morning and evening. The entire life cycle of CLM is completed in 14–50 days depending on temperature of the environment [230].

3.1.3 Damage

The CLM damages the trees by forming mines especially underside of young leaves and the fruit is rarely mined (**Figure 5**). The leaf mining by CLM results in partial chlorosis, necrosis, leaf deformation, and ultimately causes defoliation, which reduces the photosynthetic activity of plants. The mining in spring flush causes more damages as it is responsible for fruit development than that of fall flush. Besides this, mining also facilitates the entry point for number of plant pathogens such as bacterial citrus canker. The bacterium *Xanthomonas citri* (Hasse) is the cause of an important disease in citrus plant called Asian citrus canker [231]. The canker-infected tree produces lesions on leaves, stems, and fruits, which results in defoliation, fruit drop, blemishing on fruit, twig dieback, and general tree decline [232–234].

3.1.4 Management

3.1.4.1 Chemical control

Many farmers apply insecticides to mitigate the effect of CLM as they see visual impact of foliar damage on trees. This control measure is expensive and often ineffective because none of product provides long-term control than just for two week [235, 236]. To control the CLM, the chemical control is often inappropriate strategy because of high cost to manage the pest, risk of development of resistance in pest population, accumulation of pesticides residues in food and also in ground water, effect on natural control agents, harmful effect on field workers, and also to environment [237]. Tan and Huang [238] also reported development of resistance to pesticides in CLM.

Figure 4.
Adult stage of Phyllocnistis citrella *with distinct black spots at the dorsal surface of both forewing and hindwing. Photograph by Jack Kelly Clark, University of California, Statewide IPM program.*

Figure 5.
A newly developed leaf of Kinnow mandarin is under the infestation of three larvae of Phyllocnistis citrella. *Feeding of larvae upon leaf tissues results in the development of prominent mines, which is one of the diagnostic damage patterns of* P. citrella. *Photograph was taken by Dr. Muhammad Babar Shahzad Afzal at citrus nursery of Citrus Research Institute, Sargodha, Punjab, Pakistan on August 29, 2017.*

3.1.4.2 Monitoring using traps

Pheromones traps should be used to determine pest population abundance in field. However, control decision is based on sampling of active larvae. Usually, monitoring should be done in February through May and September–October, at that time 50% of trees are actively flushing. Randomly ten leaves are selected and live larvae are observed with hand lens. The young trees should be treated when 30% of leaves have active mines with live larvae, while older trees are treated unless severely damaged.

3.1.4.3 Cultural practices

In the irrigated areas of citrus production, CLM population might be suppressed by modifying the trees to produce alternate flushing by managing the irrigation and fertilizer application but it is not possible in areas where summer rainfall is too high such as Florida's subtropical climatic conditions. Other management options such as development of host plant resistance also do not provide long-term control and there is no clear evidence that varieties are resistant to attack of CLM. The water sprouts usually develop on branches and above grafting on tree trunk and rapidly produce the new flushes for long period of time. Removal of such water sprouts is crucial because it provides sites for oviposition. Besides this, the water sprouts below root stock should also be removed as they do not produce desirable fruits [239].

3.1.4.4 Biological control

The biological control agents of CLM play an important role in keeping the pest population below economic injury level. According to one report, 60% population of CLM was killed by its predators and parasitoids in Yuma. The common parasitoid *Cirrospilus coachellae* Gates (Hymenoptera: Eulophidae) appears in late summer and early fall and effectively controls the CLM. The Yuma spider mite and *Tydeu* ssp. have been reported to feed on CLM larvae [230]. Browning et al. [240] and Pena et al. [241] also identified some parasitoids of CLM {(*Cirrospilus* sp., *Pnigalio minio* (Walker), *Sympiesis* sp., *Elasmus tischeriae* (Howard), *Closterocerus cinctipennis* Ashmead, *Horismenus* sp., and *Zagrammosoma multilineatum* (Ashmead)} in Florida during 1993–1994. The level of parasitism varies ranging up to 60%. The lowest parasitism level was recorded in late winter and early spring. Similarly, some predators have also been found to attack the CLM which include *Chrysoperla rufilabris* (Burmeister), *Solenopsis invicta* (Buren), predatory thrips and some spiders [242].

3.2 Lemon butterfly

The scientific name of lemon butterfly (LBF) is *Papilio demoleus* Linnaeus and it belongs to order Lepidoptera and family Papilionidae.

3.2.1 Distribution

The rapid dispersal ability and population growth of LBF makes it economically serious pest with its distribution in different parts of the world such as Pakistan [243], Iran, Formosa, Japan, India, China [244], Bangladesh [245], United States [246], Southern Asia [247]), Island of Hispaniola [248], Iraq [249], Middle East [250, 251], Indonesia [252–254], New Guinea [255], and Australia [256, 257].

3.2.2 Eco-biology

According to Xingyong et al. [258], in warmer temperate region of China, LBF has five generations per year. This pest completes its life cycle in about over 30 days [6]. The eggs of LBF are round, having smooth surface and yellow-white in color [259]. The eggs are laid either singly or in group of 8–10 on the lower surface of leaves or twigs within 2–3 days post-mating. The hatching of eggs occurs in 4–5 days. There are five larval stages with developmental time ranging from 27 to 35 days [6]. The final larval stage is ready to form a pupa. Larvae of this stage secrete watery and soft stool as compared to grainy or dry stools produced by other stages. During the onset of pupal formation, the body of last larval instar is compressed and it produces silken thread, which is used to fasten and anchor the new developing pupa with some substratum. This stage can be called as pre-pupa. For 15–18 hours, larvae are in fixed position and shed their dried skin to form a pupa which requires 10–12 days to complete its development. When viewed from lateral part, it looks alike torpedo-shaped [260]. Both female and male adults of LBF are black with yellow or whitish marking in wings (**Figure 6**) [245]. The adults can survive 7–12 days after emergence from pupae [6]. In February (late winter), the population of LBF started to increase and reaches at peak level in April–May (late spring to early summer) and subsequently declines till September (autumn). The second peak starts at the end of September (mid-autumn) and again population flurry-up but the second peak is much smaller than the first one [245].

3.2.3 Damage

LBF is one of the major and economic pests of citrus [6, 261]. The citrus plant is damaged by only larval stage which devours the large quantity of foliage. It is voracious feeder and causes serious damage to young seedling, nurseries and also has capability to cause defoliation; thus, LBF is a potential threat to citrus growers (**Figure 7**) [262]. It can cause up to 83% defoliation in young trees in case of severe and unchecked infestation [263]. The LBF has wide ecological tolerance which enables it to thrive in different environmental conditions [264].

3.2.4 Management

3.2.4.1 Biological control

Winotai and Napompeth [265] performed a survey of natural enemies of LBF in Thailand. The *Ooencyrtus malayensis* and *Tetrastichus* sp. were found to parasitize the egg stage. There are some predators which affect the larval stage such as *Proxys punctulatus*, *Podalonia sp*, bird (*Oriolus steerii*) and spider (*Nephila sp.*) [266]. The insect parasitoids such as *Erycia nymphalidophaga* [265], *Apanteles papilionis,* and *Bracon hebetor* [267] have been reported as larval parasitoid of LBF. The parasitoids which affect the pupal stage of LBF are *Brachymeria sp* and *Pteromalus puparom*.

Figure 6.
Early instar larva (A) and adult (B) of Papilio demoleus. *Photographs by https://momsmeanderings.wordpress.com/2017/02/02/lifecycle-of-a-citrus-swallowtail-butterfly/ (A) and Foto Martein in the indoor butterfly garden "Jardins des Papillons" in Hunawihr, Alsace, France on June 22, 2007 (B).*

Figure 7.
Late instar of Papilio demoleus *feeding and cutting the leaf edges of citrus plant. Source: https://momsmeanderings.wordpress.com/2017/02/02/lifecycle-of-a-citrus-swallowtail-butterfly/.*

3.2.4.2 Insecticidal control

3.2.4.2.1 Botanical insecticides

The botanical insecticide is an alternative to synthetic insecticide for the management of pests as it provides little or no threat to the environment, ecosystem, and human health [268]. The antifeedant activity of azadirachtin was evaluated against LBF larvae [269], azadirachtin is most important natural antifeedant, which inhibits the feeding and indirectly causes death of insect due to starvation [270]. Another approach to spray the plant leaves is with distasteful substance derived from plant extracts. According to Vattikonda et al. [271], forskolin, a labdane diterpene derived from the plant *Coleus forskohlii* showed an antifeedant activity at the concentration of 200 ppm. Similarly, betulinic acid from *Ziziphus jujube* [271] and andrographolide from *Andrographis paniculate* [272] also displays strong antifeedant activity against LBF.

3.2.4.2.2 Microbial insecticides

The pathogen can be used as biological control agent and is integrated with natural enemies to protect the crop from insect pest damage. The isolates of *Isaria fumosorosea* (Ifr_1 and Ifr_2) showed their potential to kill the LBF larvae by causing 72.23 and 61.90% mortality after exposure of 8 days at dose of 10^8 spores/ml concentration, respectively [273]. The one of most commonly used microorganism is *Bacillus thuringiensis* (Bt), which is highly effective against lepidopteron larvae [274]. Osouli and Afsharmanesh [275] found that a strain of *Bacillus subtilis* (M419) produces secondary metabolites especially lipopeptide biosurfactants that inhibit the growth and cause larval mortality in LBF.

3.2.4.2.3 Synthetic insecticides

The synthetic pesticides are used in large scale for control of insect pests worldwide because of high efficacy and low cost. However, the regulation of these compounds is also increasing globally due to their concern about toxicity and environmental safety [276]. The cypermethrin has been proven more toxic for larvae of LBF than that of neem extract of *Azadirachta indica*, while deltamethrin was found less effective than *Azadirachta*; hence, cypermethrin could be a good candidate against LBF in citrus orchards [277]. Haque et al. [278] reported that LBF treated with spinosad and chlorfenapyr showed lowest rate of infestation, hence are highly effective and their use could result in highest yield of citrus due to significant suppression of LBF. The application of spinosad is environmentally benign and cost effective than that of chlorfenapyr. It can be suggested that grower should use spinosad for the management of LBF larvae in nurseries and gardens. The juvenile insect growth regulator (IGR) known as diofenolan which severely hampers the normal development, growth, and metamorphosis of LBF larvae can also be incorporated in IPM for better control [279]. Another IGR known as diflubenzuron is also effective in controlling the LBF; it not only reduces the growth but also causes morphological deformities in insect [280].

3.2.4.3 Mechanical control by hand picking

The hand picking is most practical technique under certain conditions such as in case of cheap labor availability, conspicuous, and large egg masses, when insects are too sluggish, have congregate behavior and easily accessible to the pickers. Hand picking of LBF larvae is efficient method to reduce the infestation particularly in nurseries and home gardens [281].

4. Concluding remarks

In this chapter, important features of the main citrus pests were demonstrated. Nowadays, there are many insects that are pests of citrus and a few ones were well-studied, reinforcing that other species need to be investigated. Thus, the studies on biology, ecology, and management of citrus pests are necessary, mainly those that may produce data useful to control and mitigate damages caused by the pests. In addition, field effectiveness and climatic adaptations in different geographic regions with high citrus crop activity should be evaluated in order to contribute for pest control. Furthermore, the search for environmental friendly new molecules, natural enemies, resistant plants, and technological approaches should be stimulated.

Author details

Muhammad Babar Shahzad Afzal[1,2,3*†], Ansa Banazeer[1†], Jose Eduardo Serrao[4], Muhammad Rizwan[3] and Afifa Naeem[1,5]

1 Faculty of Agricultural Sciences and Technology, Department of Entomology, Bahauddin Zakariya University, Multan, Pakistan

2 Citrus Research Institute, Sargodha, Punjab, Pakistan

3 Beekeeping and Hill Fruit Pests Research Station, Rawalpindi, Punjab, Pakistan

4 Department of General Biology, Federal University of Vicosa, Vicosa, MG, Brazil

5 Entomological Research Institute, Ayub Agriculture Research Institute, Faisalabad, Punjab, Pakistan

*Address all correspondence to: shahzad.babar35@gmail.com

† These authors contributed equally.

References

[1] Langgut D. The citrus route revealed: From Southeast Asia into the Mediterranean. Hort Science. 2017; **52**(6):814-822

[2] Noorizadeh S, Golmohammadi M, Bagheri A, Bertaccini A. Citrus industry: Phytoplasma-associated diseases and related challenges for Asia. America and Africa. Crop protection. 2022;**151**:105822

[3] Lobell DB, Field CB. Global scale climate–crop yield relationships and the impacts of recent warming. Environmental Research Letters. 2007; **2**(1):014002

[4] Tennant PF, Robinson D, Fisher L, Bennett SM, Hutton D, Coates-Beckford P, et al. Diseases and pests of citrus (Citrus spp.). Tree and Forestry Science and Biotechnology. 2009;**3**(2):81-107

[5] Tena A, Garcia-Marí F. Current situation of citrus pests and diseases in the Mediterranean basin. IOBC/WPRS Bulletin. 2011;**62**:365-368

[6] Pathak M, Rizvi Q. Effect of different temperatures and host plants on the developmental behavior of lemon butterfly, *Papilio demoleus*. Indian Journal of Entomology. 2003;**65**(4):496-499

[7] Chen XD, Stockton D, Gossett H, Qureshi JA, Ibanez F, Pelz-Stelinski KS, et al. Comparisons of economic thresholds for Asian citrus psyllid management suggest a revised approach to reduce management costs and improve yield. Frontiers in Sustainable Food Systems. 2022;**6**:365

[8] GOP Economic Threshold Level (ETL) of Different Insect pests and Diseases Directorate General of Pest Warning and Quality Control of Pesticides. Punjab, Lahore, Pakistan; 2010. Available online: https://pestwarning-agripunjab.punjab.gov.pk/economic-thresholds

[9] CABI Compendium. Dialeurodes citri (citrus whitefly). CABI International. 2022. DOI: 10.1079/cabicompendium.18698

[10] Franco JC, Suma P, Da Silva EB, Blumberg D, Mendel Z. Management strategies of mealybug pests of citrus in Mediterranean countries. Phytoparasitica. 2004;**32**(5):507-522

[11] George A, Rao CN, Mani M. Pests of citrus and their management. Trends in Horticultural Entomology. 2022:551-575

[12] Rai S, Shankar U. Integrated pest management in dry land horticultural crops. Ecologically Based Integrated Pest Management. 2012;**1**:681

[13] Sikibharathi KK, Balasubramanian P, Kumar PG. Insect pests of Khasi mandarin in east and west Khasi hills district of Meghalaya. Indian Journal of Entomology. 2022;**e21251**:1-3

[14] Kuwayama SD. Psylliden Japans. I. Transactions of the Sapporo Natural History Society. 1908;**2**:149-189

[15] Costa Lima AM. Homopteros. Insetos do Brazil. Esc. Nac. Agron. Min. Agr. 1942;**3**:1-327

[16] Capoor SP, Rao DG, Viswanath SM. *Diaphorina citri* Kuway., a vector of the greening disease of citrus in India. Indian Journal of Agricultural Sciences. 1967;**37** (572-576):1382

[17] Rao CN, Shivankar VJ, Deole S, David KJ, Dhengre VN. Insecticide resistance in field populations of Asian

citrus psyllid, *Diaphorina citri* Kuwayama (Hemiptera: Psyllidae). Pesticide Research Journal. 2014;**26**(1): 42-47

[18] Naeem A, Freed S, Jin FL, Akmal M, Mehmood M. Monitoring of insecticide resistance in *Diaphorina citri* Kuwayama (Hemiptera: Psyllidae) from citrus groves of Punjab, Pakistan. Crop Protection. 2016;**86**:62-68

[19] Grafton-Cardwell EE. How serious is the threat of Asian citrus psyllid and huanglongbing? Citrograph. 2010;**1**(1): 8-10

[20] French JV, Kahlke CJ, Da Graça JV. First record of the Asian citrus psylla, *Diaphorina citri* Kuwayama (Homoptera: Psyllidae) in Texas. Subtropical Plant Science. 2001;**53**:14-15

[21] Halbert SE, Manjunath K, Ramadugu C, Lee RF. Incidence of huanglongbing-associated '*Candidatus* Liberibacter *asiaticus*' in *Diaphorina citri* (Hemiptera: Psyllidae) collected from plants for sale in Florida. Florida Entomologist. 2012;**95**(3):617-624

[22] Bellis G, Hollis D, Jacobson S. Asian citrus psyllid, *Diaphorina citri* Kuwayama (Hemiptera: Psyllidae), and huanglongbing disease do not exist in the Stapleton Station area of the Northern Territory of Australia. Australian Journal of Entomology. 2005;**44**(1):68-70

[23] Olvera-Vargas LA, Quiroz Gaspar ÁD, Contreras-Medina DI, Aguilar-Rivera N. Potential risk analysis of Huanglongbing through geospatial technology in Colombia. Ciencia y Tecnología Agropecuaria. Dec 2020; **21**(3):e1552

[24] Halbert SE, Nunez CA. Distribution of the Asian citrus psyllid, *Diaphorina citri* Kuwayama (Rhynchota: Psyllidae) in the Caribbean Basin. Florida Entomologist. 2004;**2004**(87):401-402

[25] López-Arroyo JI, Peña MA, Rocha-Peña MA, Loera J. Ocurrencia en México del psílido asiático *Diaphorina citri* (Homóptera: Psyllidae). In: Memorias del VII Congreso Internacional de Fitopatología. Chihuahua, Chih., Méx; 2005. p. 68

[26] Subandiyah S, Nikoh N, Sato H, Wagiman F, Tsuyumu S, Fukatsu T. Isolation and characterization of two entomopathogenic fungi attacking *Diaphorina citri* (Homoptera, Psylloidea) in Indonesia. Mycoscience. 2000;**41**(5): 509-513

[27] Lashkari M, Sahragard A, Manzari S, Hosseini R, Erfanfar D. Niche modeling of Asian citrus psyllid, *Diaphorina citri* Kuwayama (Hem.: Psyllidae), in Iran. Plant Pests Research. 2013;**3**(1):45-58

[28] Rwomushana I, Khamis FM, Grout TG, Mohamed SA, Sétamou M, Borgemeister C, et al. Detection of *Diaphorina citri* Kuwayama (Hemiptera: Liviidae) in Kenya and potential implication for the spread of Huanglongbing disease in East Africa. Biological Invasions. 2017;**19**(10): 2777-2787

[29] Nakata T. Temperature-dependent development of the citrus psyllid, *Diaphorina citri* (Homoptera: Psylloidea), and the predicted limit of its spread based on overwintering in the nymphal stage in temperate regions of Japan. Applied Entomology and Zoology. 2006;**41**(3):383-387

[30] Al-Zadjali TS, Ibrahim R, Al-Rawahi AK. 0039. First record of *Diaphorina citri* Kuwayama (Hemiptera: Psyllidae) from the Sultanate of Oman. Insecta Mundi. 2008;**0039**:1-3

[31] De Bac G, Saponari M, Loconsole G, Martelli GP, Yokomi RK, Catalano L, et al. First report of "*Candidatus* Liberibacter *asiaticus*" associated with Huanglongbing in Ethiopia. Plant Disease. 2010;**94**(4):482

[32] Lim WH, Shamsudin OM, Ko WW. Citrus greening and alternate hosts of the vector, *Diaphorina citri* Kuw., in P. Malaysia. MAPPS Newsletter. 1990; **13**(4):56-58

[33] Catling HD, Garnier M, Bové JM. Presence of citrus greening disease in Bangladesh and a new method for rapid diagnosis. FAO Plant Protection Bulletin. 1978;**26**(1):16-18

[34] Hall DG, Albrigo LG. Estimating the relative abundance of flush shoots in citrus with implications on monitoring insects associated with flush. HortScience. 2007;**42**:364-368

[35] Husain MA, Nath D. The citrus psylla (*Diaphorina citri*, Kuw.) [Psyllidae: Homoptera]. Memoirs of the Department of Agriculture in India. Entomological Series. 1927;**10**(2):5-27

[36] Qureshi JA, Rogers ME, Hall DG, Stansly PA. Incidence of invasive *Diaphorina citri* (Hemiptera: Psyllidae) and its introduced parasitoid *Tamarixia radiata* (hymenoptera: Eulophidae) in Florida citrus. Journal of Economic Entomology. 2009;**102**:247-256

[37] Tsai JH, Liu YH. Biology of *Diaphorina citri* (Homoptera: Psyllidae) on four host plants. Journal of Economic Entomology. 2000;**93**:1721-1725

[38] Fung YC, Chen CN. Effects of temperature and host plant on population parameters of the citrus psylla (*Diaphorina citri* Kuwayama). Formosan Entomology. 2006;**26**:109-123

[39] Chen X, Liao C. Studies on the biology of *Diaphorina citri* and its relationship with HLB. China Citrus. 1982;**4**:4-17

[40] Bindra OS. Insect pests of citrus, review of work done, lacunae in our knowledge and future line of work. In: ICAR Workshop Fr. Res. Ludhiana: Panjab Agricultural University; 1969. pp. 128-130

[41] Pande YD. Biology of citrus psylla, *Diaphorina citri* Kuw. (Hemiptera: Psyllidae). Israel Journal of Entomology. 1971;**6**(2):307-311

[42] Iqbal J, Hussain HN, Latif M, Baig MB, Owayss AA, Raweh HS, et al. A field study investigating the insecticidal efficacy against *Diaphorina citri* Kuwayama on Kinnow mandarin, *Citrus reticulata* Blanco trees. Saudi Journal of Biological Sciences. 2020;**27**(5): 1237-1241

[43] Michaud JP. Natural mortality of Asian citrus psyllid (Homoptera: Psyllidae) in Central Florida. Biological Control. 2004;**29**:260-269

[44] Gravena S. Practical Handbook of Ecological management of Citrus pest. Jaboticabal, SP, Brazil: Gravena; 2005

[45] Walter AJ, Hall DG, Duan YP. Low incidence of '*Candidatus* Liberibacter *asiaticus*' in *Murraya paniculata* and associated *Diaphorina citri*. Plant Disease. 2012;**96**:827-832

[46] Halbert S, Niblett CL, Manjunath KL, Lee RF, Brown LG. Establishment of two new vectors of citrus pathogens in Florida. Proceedings of the International Society of Citriculture IX Congress. 2000: 1016-1017

[47] Grafton-Cardwell EE, Stelinski LL, Stansly PA. Biology and management of Asian citrus psyllid, vector of the huanglongbing pathogens. Annual Review of Entomology. 2013;**7**(58): 413-432

[48] Boina DR, Bloomquist JR. Chemical control of the Asian citrus psyllid and of huanglongbing disease in citrus. Pest Management Science. 2015;**71**(6): 808-823

[49] Bove JM. Huanglongbing: A destructive, newly-emerging, century-old disease of citrus. Journal of Plant Pathology. 2006;**88**:7-37

[50] Quarles W. IPM for Asian citrus psyllid and huanglongbing disease. IPM Practitioner. 2013;**34**:1-7

[51] Kuchment A. The End of Orange Juice. New York: Scientific American; 2013. pp. 52-59

[52] Halbert SE, Manjunath KL. Asian citrus psyllids (Sternorrhyncha: Psyllidae) and greening disease of citrus: A literature review and assessment of risk in Florida. Florida Entomologist. 2004;**87**:330-353

[53] Bassanezi RB, Montesino LH, Stuchi ES. Effects of Huanglongbing on fruit quality of sweet orange cultivars in Brazil. European Journal of Plant Pathology. 2009;**125**:565-572

[54] Baldwin E, Plotto A, Manthey J, McCollum G, Bai J, Irey M, et al. Effect of Liberibacter infection (Huanglongbing disease) of citrus on orange fruit physiology and fruit/fruit juice quality: Chemical and physical analyses. Journal of Agricultural and Food Chemistry. 2010;**58**:1247-1262

[55] Graham JH, Johnson EG, Gottwald TR, Irey MS. Presymptomatic fibrous root decline in citrus trees caused by huanglongbing and potential interaction with Phytophthora spp. Plant Disease. 2013;**97**(9):1195-1199

[56] Johnson EG, Wu J, Bright DB, Graham JH. Association of '*Candidatus* Liberibacter *asiaticus*' root infection, but not phloem plugging with root loss on huanglongbing-affected trees prior to appearance of foliar symptoms. Plant Pathology. 2014;**63**:290-298

[57] Yang Y, Huang M, Andrew G, Beattie C, Xia Y, Ouyang G, et al. Distribution, biology, ecology and control of the psyllid *Diaphorina citri* Kuwayama, a major pest of citrus: A status report for China. International Journal of Pest Management. 2006; **2006**(52):343-352

[58] Michaud JP. Biological control of Asian citrus psyllid, *Diaphorina citri* (Homoptera: Psyllidae), in Florida: A preliminary report. Entomological News. 2002;**113**:216-222

[59] Batra RC, Sharma DR, Singh RA, Singh SN. Status of citrus insect and mite pests and their natural enemies in different agro-climatic zones of Punjab. Indian Journal of Horticulture. 1990; **47**(3):331-336

[60] Aubert B. *Trioza erytreae* del Guercio and *Diaphorina citri* Kuwayama (Homoptera: Psyllidae), the two vectors of citrus greening disease: Biological aspects and possible control strategies. Fruits. 1987;**42**:149-162

[61] Chien CC. The role of parasitoids in the pest management of citrus psyllid. In: Proceedings of the Symposium on Research and Development of Citrus in Taiwan, Taichung, Taiwan. 1995. pp. 245-261

[62] Shafee SA, Alam SM, Agarwal MM. Taxonomic Survey of Encyrtid Parasites

(Hymenoptera: Encyrtidae) in India. Vol. 10. India: Aligarh Muslim University; 1975. p. 125

[63] McFarland CD, Hoy MA. Survival of *Diaphorina citri* (Homoptera: Psyllidae), and its two parasitoids, *Tamarixia radiate* (hymenoptera: Eulophidae) and *Diaphorencyrtus aligarhensis* (hymenoptera: Encyrtidae), under different relative humidities and temperature regimes. Florida Entomologist. 2001;**84**(2):227-233

[64] Skelley LH, Hoy MA. A synchronous rearing method for the Asian citrus psyllid and its parasitoids in quarantine. Biological Control. 2004;**29**:14-23

[65] Belasque J Jr, Bassanezi RB, Yamamoto PT, Ayres AJ, Tachibana A, Violante AR, et al. Lessons from huanglongbing management in São Paulo state, Brazil. Journal of Plant Pathology. 2010;**92**:285-302

[66] Stansly PA, Rogers ME. Managing Asian citrus psyllid populations. Citrus Ind. 2006;**87**:1-2

[67] Qureshi JA, Kostyk BC, Stansly PA. Effectiveness of selective insecticides to control Asian citrus psyllid and citrus leafminer during leaf flushing. Proceedings of the Florida State Horticultural Society. 2011;**124**:85-89

[68] Rogers ME, Stansly PA, Stelinski LL. Florida citrus pest management guide: Asian citrus psyllid and citrus leafminer. In: IFAS Extension Publication ENY 734. Gainesville, FL: Institute of Food and Agricultural Sciences, University of Florida; 2014

[69] Leong SC, Abang TF, Beattie A, Kueh RHJ, Wong SK. Impacts of horticultural mineral oils and two insecticide practices on population fluctuation of *Diaphorina citri* and spread of huanglongbing in a citrus orchard of Sarawak. The Scientific World Journal. 2012;**2012**:1-7

[70] Aubert B. Bove JM, Effect of penicillin or tetracycline injections of citrus trees affected by greening disease in Reunion Island. In: Calavan EC, Garnsey SM, Timmer LW, editors. Proceedings of the 8th Conference of the International Organization of Citrus Virologists, Riverside, CA. Riverside, CA: The International Organization of Citrus Virologists; 1980. pp. 103-108

[71] Bellows TS, Meisenbacher C. Field population biology of citrus whitefly, *Dialeurodes citri* (Ashmead) (Heteroptera: Aleyrodidae), on oranges in California. Population Ecology. 2007; **49**(2):127-134

[72] Ashmead WH. The orange Aleurodes (*Aleurodes citrin*. sp.). Florida Dispatch. 1885;**2**(42):704

[73] Woodworth CW. Whitefly eradication. In: University of California Agricultural Experiment Station, Circular No. 32. Berkeley: University of California; 1907. p. 15

[74] Chen C, Klein ZI, Shabtai Y. A new citrus pest in Israel, the citrus whitefly, *Dialeurodes citri* Ashm.Prevention of its expansion and the eradication program. Alon haNotea. 1979;**33**:673-680

[75] Argov Y, Rössler Y, Voet H, Rosen D. The biology and phenology of the citrus whitefly, *Dialeurodes citri*, on citrus in the coastal plain of Israel. Entomologia Experimentalis et Applicata. 1999;**93**(1): 21-27

[76] Uygun N, Ulusoy MR, Sekeroglu E, Ohnesorge B, Gozel U. Interactions between two introduced species of whiteflies in the Mediterranean area of Turkey: *Dialeurodes citri* (Ashmead)

and Parabemisia myricae (Kuwana) (Hom., Aleyrodidae). Journal of Applied Entomology. 1994;**118**: 365-369

[77] Yigit A, Canhilal R, Ekmekci U. Seasonal population fluctuations of *Serangium parcesetosum* (Coleoptera: Coccinellidae), a predator of citrus whitefly, *Dialeurodes citri* (Homoptera: Aleyrodidae) in Turkey's eastern Mediterranean citrus groves. Environmental Entomology. 2003;**32**: 1105-1114

[78] Žanić K, Igrc Barčić J, Kačić S. *Dialeurodes citri* (Ashmead, 1885) in the Adriatic region of Croatia. Agriculturae Conspectus Scientificus. 2001;**66**(3): 161-168

[79] Gill G. Citrus whitefly blows in. N Z Biosecurity. 2001;**26**:19

[80] Tayyib M, Mukhtar M, Yousuf MJ, Akbar M. Contribution to the study of genus Dialeurodes (Homoptera: Aleyrodidae) from Pakistan. Pakistan Entomologist. 2019;**41**(1):35-38

[81] Mound LA, Halsey SH. Whitefly of the World. British Museum (Natural History). Chichester: John Wiley & Sons; 1978. p. 340

[82] Wang P, Song X, Zhang H. Isolation and characterization of *Aschersonia placenta* from citrus orchards and its pathogenicity towards *Dialeurodes citri* (Ashmead). Journal of Invertebrate Pathology. 2013;**112**(2): 122-128

[83] Sulaymonov OA, Dusmurodova GT, Sobirov BB. Studying the efficacy of allchungkill cyc.k. against citrus whitefly (*Dialeurodes citri* ashm.) in culture lemon. The American Journal of Agriculture and Biomedical Engineering. 2022;**4**(05):22-25

[84] Kumar R, Batra RC, Sharma DR. Association of sooty mould development with citrus whitefly, *Dialeurodes citri* Asmead, infestation on different citrus rootstocks. Integrated plant. Disease Management. 2011:71

[85] Vinay S, Galjrav S, Kumar SA. Vertical distribution and abundance of aleyrodids of agricultural importance in Himachal Pradesh. Indian Journal of Entomology. 2021;**83**:551-557

[86] Fasulo TR, Weems HV. Citrus whitefly, *Dialeurodes citri* (Ashmead) (Insecta: Hemiptera: Aleyrodidae). UF/ IFAS Extension. 1999;**1999**(128):1-6

[87] Porath A. Whiteflies (Aleurodidae) Attacking Citrus in the World. Israel Ministry of Agriculture Extension Service, Citriculture Department; 1979. p. 28 (Mimeograph, in Hebrew)

[88] Barbagallo S, Longo S, Rapisarda C. White-flies and psyllids injurious to citrus. In: Integrated Pest Control in Citrus-Groves. Boca Raton, Florida, United States: CRC Press; 1986. pp. 89-98

[89] Zhang YH, Liu CH, Wang Q, Wang YL, Zhou CY, Zhou Y. Identification of *Dialeurodes citri* as a vector of citrus yellow vein clearing virus in China. Plant Disease. 2019, 2019; **103**(1):65-68

[90] Catara A, Azzaro A, Davino M, Polizzi G. Yellow vein clearing of lemon in Pakistan. In: International Organization of Citrus Virologists Conference Proceedings (1957-2010). Vol. 12(12). USA: The Regents of the University of California; 1993

[91] Zhou Y, Chen HM, Cao MJ, Wang XF, Jin X, Liu KH, et al. Occurrence, distribution and molecular

characterization of Citrus yellow vein clearing virus in China. Plant Disease. 2017;**101**:137-143

[92] Loconsole G, Önelge N, Potere O, Giampetruzzi A, Bozan O, Satar S, et al. Identification and characterization of citrus yellow vein clearing virus, a putative new member of the genus Mandarivirus. Phytopathology. 2012; **102**:1168-1175

[93] Chen HM, Li ZA, Wang XF, Zhou Y, Tang KZ, Zhou CY, et al. First report of Citrus yellow vein clearing virus on lemon in Yunnan, China. Plant Disease. 2014;**98**(12):1747

[94] Kaneko S. Seasonal and yearly change in adult abundance of a predacious ladybird *Serangium japonicum* (Coleoptera: Coccinellidae) and the citrus whitefly *Dialeurodes citri* (Hemiptera: Aleyrodidae) in citrus groves. Applied Entomology and Zoology. 2017;**52**:481-489

[95] Yiğit A, Canhilal R. Establishment and dispersal of *Serangium parcesetosum* Sicard (Coleoptera, Coccinellidae), a predatory beetle of citrus whitefly, *Dialeurodes citri* Ashm.(Homoptera, Aleyrodidae) in the East Mediterranean region of Turkey/Die Etablierung und Verbreitung von *Serangium parcesetosum* Sicard (Coleoptera, Coccinellidae), ein Prädator der Weissen Fliege der Citrusgewächse, *Dialeurodes citri* Ashm. (Homoptera, Aleyrodidae) im Ostmediterranen Gebiet der Türkei. Zeitschrift für Pflanzenkrankheiten und Pflanzenschutz/Journal of Plant Diseases and Protection. 2005;**112**: 268-275

[96] Hudson WG, Williams ML. Release of the parasitic wasp, *Encarsia lahorensis* (hymenoptera: Aphelinidae), for control of citrus whitefly (Homoptera: Aleyrodidae) on gardenia in Alabama. Environmental Entomology. 1986;**15**(3): 585-589

[97] Ulusoy MR, Uygun N, Kersting U, Karaca I, Satar S. Present status of citrus whiteflies (Homoptera: Aleyrodidae) in Turkey and their control/Weiße Fliegen (Homoptera: Aleyrodidae) an Zitrus und deren Bekämpfung in der Türkei. Zeitschrift für Pflanzenkrankheiten und Pflanzenschutz/Journal of Plant Diseases and Protection. 1996;**103**:397-402

[98] Fadamiro HY, Xiao YF, Hargroder T, Nesbitt M, Umeh V, Childers CC. Seasonal occurrence of key arthropod pests and associated natural enemies in Alabama Satsuma citrus. Environmental Entomology. 2008;**37**:555-567

[99] Gao RX, Ou YZ, Gao ZX, Zheng JX. A preliminary report on *Aschersonia aleyrodis* to control citrus whitefly. Journal of Fujian Agricultural College. 1985;**14**:127-133

[100] Yu S, Ding L, Luo R, Li X, Yang J, Liu H, et al. Identification of immunity-related genes in *Dialeurodes citri* against entomopathogenic fungus *Lecanicillium attenuatum* by RNA-Seq analysis. PLoS One. 2016; **11**(9):e0162659

[101] Yu SJ, Pan Q, Luo R, Wang CL, Cheng LY, Yang JS, et al. Expression of exogenous dsRNA by *Lecanicillium attenuatum* enhances its virulence to *Dialeurodes citri*. Pest Management Science. 2019;**75**(4):1014-1023

[102] Doukas D, Payne CC. Greenhouse whitefly (Homoptera: Aleyrodidae) dispersal under different UV-light environments. Journal of Economic Entomology. 2007;**100**:389-397

[103] Honda K. Reactions to light in insects and practical applications.

Journal of Biomechanics. 2011;**35**:233-236

[104] Johansen NS, Vänninen I, Pinto DM, Nissinen AI, Shipp L. In the light of new greenhouse technologies: 2. Direct effects of artificial lighting on arthropods and integrated pest management in greenhouse crops. Annals of Applied Biology. 2011;**159**(1):1-27

[105] Ben-Yakir D, Antignus Y, Offir Y, Shahak Y. Optical manipulations: An advance approach for reducing sucking insect pests. In: Advanced Technologies for Managing Insect Pests. Dordrecht: Springer; 2013. pp. 249-267

[106] Emura K, Tazawa S. The development of the eco-engineering insect control technology-physical control of insect behavior using artificial lights. Eco-engineering. 2004;**16**:237-240

[107] Tariq K, Noor M, Saeed S, Zhang H. The effect of ultraviolet-a radiation exposure on the reproductive ability, longevity, and development of the *Dialeurodes citri* (Homoptera: Aleyrodidae) F1 generation. Environmental Entomology. 2015;**44**(6):1614-1618

[108] Gunn A. The determination of larval phase coloration in the African armyworm, *Spodoptera exempta* and its consequences for thermoregulation and protection from UV light. Entomologia Experimentalis et Applicata. 1998;**1998**(86):125-133

[109] Meng JY, Zhang CY, Zhu F, Wang XP, Lei CL. Ultraviolet light-induced oxidative stress: Effects on antioxidant response of *Helicoverpa armigera* adults. Journal of Insect Physiology. 2009;**55**(6):588-592

[110] Cowan T, Gries G. Ultraviolet and violet light: Attractive orientation cues for the Indian meal moth. *Plodia interpunctella*. Entomologia Experimentalis et Applicata. 2009;**131**(2):148-158

[111] Byrne DN, Von Bretzel PK, Hoffman CJ. Impact of trap design and placement when monitoring for the banded winged whitefly and the sweetpotato whitefly (Homoptera: Aleyrodidae). Environmental Entomology. 1986;**15**(2):300-304

[112] Tariq K, Ali A, Naz G, Anwar Z, Khan AA, Jan I, et al. Field evaluation of yellow sticky traps for the control of citrus whitefly *Dialeurodes citri* (Ashmead) (Homoptera: Aleyrodidae) in citrus orchard of Pakistan. World Journal of Zoology. 2016;**11**(2):110-116

[113] Uygun N, Ohnesorge B, Ulusoy R. Two species of whiteflies on citrus in eastern Mediterranean: *Parabemisia myricae* (Kuwana) and *Dialeurodes citri* (Ashmead) morphology, biology, host plants and control in southern Turkey. Journal of Applied Entomology. 1990;**110**(1-5):471-482

[114] Uygun N, Satar S. The current situation of citrus pests and their control methods in Turkey. IOBC-WPRS Bulletin. 2008;**38**:2-9

[115] Gokkes M, Tadmor U, Carmon D, Shneider R. An uncorroborated field study for the control of citrus white fly, *Dialeurodes citri* (Ashmead). Alon Hanotea. 1990;**44**(8):759-761

[116] Tekchand S, Jaglan MS, Yadav SS, Yadav M. Bio-efficacy of different insecticides in the management of citrus whitefly, *Dialeurodes citri* (Ashmead) on citrus mandarin (var. Kinnow). Pesticide Research Journal. 2016;**28**(2):231-236

[117] Gupta BP, Joshi KC. Evaluation of some insecticides for the control of the Citrus White Fly (*Dialeurodes Citri* Ashmead). Indian Journal of Horticulture. 1977;**34**(1):34-39

[118] Horowitz AR, Ishaaya I. Chemical control of Bemisia-management and application. Bemisia. 1995:537-556

[119] Kunimi Y. Current status and prospects on microbial control in Japan. Journal of Invertebrate Pathology. 2007; **95**:181-186

[120] Ahmed NH, Abd-Rabou SM. Host plants, geographical distribution, natural enemies and biological studies of the citrus mealybug, *Planococcus citri* (Risso) (Hemiptera: Pseudococcidae). Egyptian Academic Journal of Biological Sciences. A, Entomology. 2010;**3**(1): 39-47

[121] Gill HK, Goyal G, Gillett-Kaufman J. Citrus Mealybug *Planococcus citri* (Risso) (Insecta: Hemiptera: Pseudococcidae). EDIS. 2012;**2012**(9): 1-4

[122] Bartlett BR, Lloyd DC. Mealybugs attacking citrus in California—A survey of their natural enemies and the release of new parasites and predators. Journal of Economic Entomology. 1958;**51**(1): 90-93

[123] Seabra SG, Brás PG, Zina V, Borges da Silva E, Rebelo MT, Figueiredo E, et al. Molecular evidence of polyandry in the citrus mealybug, *Planococcus citri* (Hemiptera: Pseudococcidae). PLoS One. 2013;**8**(7):e68241

[124] Kütük H, Karacaoglu M, Tüfekli M, Satar G, Yarpuzlu F. Study on field evolution of citrus mealybug (*Planococcus citri* Risso) (Hemiptera: Pseudococcidae) management in Finike county of Antalya, Turkey. Journal of the Entomological Research Society. 2014; **16**(3):101-107

[125] Niekerk SV, Malan AP. Citrus mealybug highly susceptible to entomopathogenic nematodes, with humidity being the main impediment for successful aerial application. SA Fruit Journal. 2012;**11**(5):49-53

[126] Radosevich DL, Cloyd RA. Spray volume and frequency impacts on insecticide efficacy against the citrus mealybug (Hemiptera: Pseudococcidae) on coleus under greenhouse conditions. Journal of Entomological Science. 2021; **56**(3):305-320

[127] Williams DJ, Watson GW. Scale Insects of the Tropical South Pacific Region. Part 2. Mealybugs (Pseudococcidae). Wallingford, Oxon, UK: CAB International; 1988. p. 260

[128] Smith D, Beattie GA, Broadley R. Citrus Pests and their Natural Enemies: Integrated Pest Management in Australia. Brisbane, Australia: Queensland DPI & HRDC; 1997

[129] Williams DJ. Mealybugs of Southern Asia. Natural History Museum/ Southdene Sdn Bhd; 2004

[130] Gupta RK, Kour R, Gani M, Guroo MA, Bali K. Potential of wax degrading bacteria for management of the citrus mealybug, *Planococcus citri*. BioControl. 2022;**67**(1):49-61

[131] Cox JM. The mealybug genus *Planococcus* (Homoptera: Pseudococcidae). Bulletin of the British Museum (Natural History), Entomology. 1989;**58**(1):1-78C

[132] Blumberg D, Klein M, Mendel Z. Response by encapsulation of four mealybug species (Homoptera: Pseudococcidae) to parasitization by

Anagyrus pseudococci. Phytoparasitica. 1995;**23**(2):157-163

[133] Wakgari WM, Giliomee JH. The biology of three mealybug species (Hemiptera: Pseudococcidae) found on citrus in the Western Cape Province, South Africa. African Entomology. 2003; **11**:173-182

[134] Arshad M, Majeed MZ, Ullah MI, Ahmad K, Tayyab M, Yahya M. Laboratory evaluation of some insecticides against citrus mealybug *Planococcus citri* (Homoptera: Pseudococcidae). Journal of Entomology and Zoology Studies. 2015;**2015**(3):20-23

[135] Mahmood R, Rehman A, Ahmad M. Prospects of biological control of citrus insect pests in Pakistan. Journal of Agricultural Research. 2014;**52**(2): 229-244

[136] McKenzie HL. Mealybugs of California: With Taxonomy, Biology and Control of North American Species (Homoptera: Coccoidea: Pseudococcidae). Berkeley, CA: University of California Press; 1967

[137] Williams DJ, Willink MCG. Mealybugs of Central and South America. Wallingford, UK: CAB International; 1992

[138] Bindra OS. Citrus Decline in India. Ludhiana, Punjab: Jt. Publi. of PAU/ OSU/USAID; 1970

[139] Clausen CP. Mealybugs of citrus trees. California Agricultural Experimental Station Bulletin. 1915;**258**: 19-48

[140] Uygun N, Ulusoy MR, Karaca İ, Satar S. Meyve ve bağ zararlıları. Özyurt Matbaacılık: Adana; 2010. p. 347

[141] Franco JC, Zada A, Mendel Z. Novel approaches for the management of mealybug pests. In: Horowitz IIAR, editor. Biorational Control of Arthropod Pests: Application and Resistance management. Dordrecht: Springer Verlag Netherlands; 2009. pp. 233-278

[142] Waterworth RA, Wright IM, Millar JG. Reproductive biology of three cosmopolitan mealybug (Hemiptera: Pseudococcidae) species, *Pseudococcus longispinus*, *Pseudococcus viburni*, and *Planococcus ficus*. Annals of the Entomological Society of America. 2011; **104**(2):249-260

[143] Thornhill R, Alcock J. The evolution of insect mating systems. In: The Evolution of Insect Mating Systems. Cambridge, Mass: Harvard University Press; 2013

[144] Hattingh V, Tate BA. The pest status of mealybugs on citrus in southern Africa. Proceedings of the International Society of Citriculture. 1996;**1**:560-563

[145] Hattingh V. Mealybugs and cottony cushion scale on citrus in southern Africa. Citrus Journal. 1993;**3**:2022

[146] Kerns DL. Integrated pest management of citrus mealybug. 2004

[147] Hattingh V, Moore SD. Mealybugs. In: Grout TG, editor. integrated Production Guidelines for Export Citrus: Integrated pest and Disease management. Vol. 3. Nelspruit: Citrus Research International; 2003. pp. 65-69

[148] Pérez-Rodríguez J, Pekas A, Tena A, Wäckers FL. Sugar provisioning for ants enhances biological control of mealybugs in citrus. Biological Control. 2021;**157**:104573

[149] Dreishpun Y. A guide to the control of citrus pests. Ministry of Agriculture

and Rural Development, Bet Dagan, Israel. 2000. Publ. No. 17003.

[150] Shivankar VJ, Singh S. Citrus Insect Pests. Nagpur: NRCC Publication; 2000. p. 252

[151] Pathak KA et al. Proceedings of the International Symposium of Citriculture, Held at Nagpur during Nov. 23-27th, Nagpur, India: Indian Society of Citriculture; 1999. pp. 956-960

[152] Singh SP. Some Success Stories in Classical Biological Control of Agricultural Pests in India. Bangkok (Thailand) APAARI/FAORAP: Apaari Publication; 2004. p. 21

[153] Al A, El-Arnaouty SA, Attia AR, Abd-Alla-Ael M. Biological control of citrus mealybug, *Planococcus citri* (Risso) using coccinellid predator, *Cryptolaemus montrouzieri*. Muls. Pakistan Journal of Biological Science. 2010;**13**(5):216-222

[154] Hattingh V, Tate BA. Effects of field-weathered residues of insect growth regulators on some Coccinellidae (Coleoptera) of economic importance as biological control agents. Bulletin of Entomological Research. 1995;**85**: 489-493

[155] Kerns D, Wright G, Loghry J. Citrus mealybug (*Planococcus citri*). Citrus Arthropod Pest Management in Arizona. 2004

[156] Afifi AI, El Arnaouty SA, Attia AR. Biological control of citrus mealybug, *Planococcus citri* (Risso) using coccinellid predator, *Cryptolaemus montrouzieri* Muls. Pakistan Journal of Biological Sciences: PJBS. 2010;**13**(5):216-222

[157] De Jong PW, Van Alphen JJM. Host size selection and sex allocation in *Leptomastix dactylopii*, a parasitoid of *Planococcus citri*. Entomologia Experimentalis et Applicata. 1989;**50**: 161-169

[158] Cadée N, van Alphen JJ. Host selection and sex allocation in *Leptomastidea abnormis*, a parasitoid of the citrus mealybug *Planococcus citri*. Entomologia Experimentalis et Applicata. 1997;**83**(3):277-284

[159] Mahfoudhi N, Dhouibi MH. Survey of mealybugs (Hemiptera: Pseudococcidae) and their natural enemies in Tunisian vineyards. African Entomology. 2009;**17**(2):154-160

[160] Mani M. Effectiveness of the exotic encyrtid parasitoid, *Leptomastix dactylopii* How. In the control of mealybug, *Planococcus citri* (Risso) in guava orchards. Journal of Entomological Research. 1994;**18**(4): 351-355

[161] Krishnamoorthy A, Singh SP. Biological control of citrus mealybug, *Planococcus citri* with an introduced parasite, *Leptomastix dactylopii* in India. Entomophaga. 1987;**32**(2): 143-148

[162] Hefetz A, Tauber O. Male response to the synthetic sex pheromone of *Planococcus citri* (Risso) (Hom., Pseudococcidae) and its application for population monitoring. Journal of Applied Entomology. 1990;**109**(1-5): 502-506

[163] Kerns D, Wright GC, Loghry J. Citrus Arthropod Pest Management. Phoenix, AZ: Arizona Department of Agriculture; 2002. p. 29

[164] Öztop A, Koçlu T, Elekçioğlu N, Karacaoğlu M, Tekşam İ, Ölçülü M, et al. Turunçgil Bahçelerinde Entegre Mücadele Teknik Talimatı. Ankara, Turkey: Ministry of

Agriculture and Rural Affairs, General Directorate of Agricultural Research; 2011. p. 171

[165] de Souza JC, Reis PR, Ribeiro JA, Santa Cecília LVC, Silva RA. Chemical control of the coffee root mealybug *Dysmicoccus texensis* (Tinsley, 1900) in coffee plants (*Coffea arabica* L). Coffee Science. 2007;**2**:29-37

[166] Bailey SF. Thrips of economic importance in California. California Agricultural Experiment Station Circular. 1938;**346**:1-77

[167] Ananthakrishnan TN. Thrips Biology and Control. New Delhi: Macmillan; 1973. p. 120

[168] Ananthakrishnan TN. Bionomics of thrips. Annual Review of Entomology. 1993;**38**(1):71-92

[169] Blank RH, Gill GS. Thrips (Thysanoptera: Terebrantia) on flowers and fruit of citrus in New Zealand. New Zealand Journal of Crop and Horticultural Science. 1997;**25**(4): 319-332

[170] Mound LA, Jackman DJ. Thrips in the economy and ecology of Australia. In: Zalucki MP, Drew RAI, White GG, editors. Proceedings of the Sixth Australian Applied Entomological Research Conference. Vol. 1. St Lucia, Australia: University of Queensland; 1998. pp. 472-478

[171] Vassiliou VA. Bio-ecology and control of *Pezothrips kellyanus* (Thysanoptera: Thripidae) on citrus in Cyprus. European Journal of Plant Science and Biotechnology. 2008;**2**:69-73

[172] Trabelsi I, Boulahia-Kheder S. Sur la présence en Tunisie du thrips des agrumes *Pezothrips kellyanus* (Thysanoptera: Thripidae). Annales de l'INRAT. 2009;**82**:181-186

[173] Marullo R, De Grazia A. Thripidae. In: Vacante V, Gerson U, editors. Integrated Control of Citrus Pests in the Mediterranean Region. 1st ed. Bentham Science: Sharjah, United Arab Emirates; 2012. pp. 109-118

[174] Mueller TG, Kahl HM, Cass BN, Grafton-Cardwell EE, Rosenheim JA. Differential impacts of citrus thrips across sweet orange and mandarin species. Journal of Economic Entomology. 2019;**112**(6):2767-2773

[175] Belaam-Kort I, Mansour R, Attia S, Boulahia KS. Thrips (Thysanoptera: Thripidae) in northern Tunisian citrus orchards: Population density, damage and insecticide trial for sustainable pest management. Phytoparasitica. 2021; **49**(4):527-538

[176] Horton JR. The citrus thrips. USDA Bulletin 616. 1918

[177] Kerns D, Wright G, Loghry J. Citrus Thrips (*Scirtothrips citri*). The University of Arizona Citrus Insects Cooperative Extension. 2001. Available from: http://cals.arizona.edu/crops/citrus/insects/citrusinsect.html [Accessed: 13 December 2017]

[178] Diffie S, Edwards GB, Mound LA. Thysanoptera of Southeastern U.S.A: A checklist for Florida and Georgia. Zootaxa. 2008;**1787**:45-62

[179] Stuart RR, Gao YL, Lei ZR. Thrips: Pests of concern to China and the United States. Agricultural Sciences in China. 2011;**10**(6):867-892

[180] EFSA Panel on Plant Health (PLH), Jeger M, Bragard C, Caffier D, Candresse T, Chatzivassiliou E, et al. Pest categorisation of *Scirtothrips citri*. EFSA Journal. 2018;**16**(3):e05189

[181] Bagnall RS. Brief descriptions of new Thysanoptera VII. The Annals and

Magazine of Natural History. 1916;**17**: 213-223

[182] Mound LA, Walker AK. Terebrantia (Insecta: Thysanoptera). Fauna of New Zealand. 1982;**1**:120

[183] Moritz G, Mound LA, Morris DC, Goldarazena A. Pest Thrips of the World: Visual and Molecular Identification of pest Thrips. CD-Rom. Brisbane, Australia: CBIT; 2004

[184] Marullo R. *Pezothrips kellyanus* unnuovo thripide parassita delle culture meridionali. Informatore Fitopatologico. 1998;**10**:72-74

[185] Conti F, Tumminelli R, Amico C, Fisicaro R, Frittitta C, Perrotta G, et al. Monitoring *Pezothrips kellyanus* on citrus in eastern Sicily. Thrips and Tospoviruses. In: Proceedings of the 7th International Symposium on Thysanoptera, Reggio Calabria, Italy. 2001. pp. 207-210

[186] Navarro-Campos C, Aguilar A, Garcia-Marí F. Population trend and fruit damage of *Pezothrips kellyanus* in citrus orchards in Valencia (Spain). Integrated control in citrus fruit crops. IOBC/WPRS Bulletin. 2011;**62**:285-292

[187] Palmer JM. *Megalurothrips* in the flowers of tropical legumes: A morphometric study. In: Holman J, Pelikan J, Dixon AFG, Weismann L, editors. Population Structure, Genetics and Taxonomy of Aphids and Thysanoptera. The Hague, Netherlands: SPB Academic Publishing; 1987. pp. 480-495

[188] Varikou K, Tsitsipis JA, Alexandrakis V, Mound LA. *Pezothrips kellyanus* (Bagnall) (Thysanoptera: Thripidae), a new pest of citrus trees in Crete. In: Proceedings of the VII European Congress of Entomology, DTP Art Work Thessaloniki, Greece, 7–13 October 2002. p. 33

[189] Costa L, Mateus C, zur Strassen R, Franco J. Thrips (Thysanoptera) associated to lemon orchards in the Oeste region of Portugal. IOBC/WPRS Bulletin. 2006;**29**:285-291

[190] [CNPPO] Chilean National Plant Protection Organization. Chile Pest List. 2007. Available from: http://www.aqis.gov.au/PhytoPublish/Restrictions/Chile%20pest%20list.doc

[191] Abbassi M. Les thrips des agrumes. Données préliminaires. Agriculture du Maghreb N°. 2014;**76**:32

[192] Andrewartha HG. Introduction to the Study of Animal Populations. Chicago: University Chicago Press; 1971. p. 262

[193] Munger F. Notes on the biology of the citrus thrips. Journal of Economic Entomology. 1942;**35**:455

[194] Lewis T. Thrips, their biology, ecology and economic importance. Thrips, their biology, ecology and economic importance. 1973;**49**:xv+349

[195] Tanigoshi LK, Nishio-Wong JY. Citrus thrips: Biology, ecology, and control. US Department of Agriculture Technical Bulletin. 1982;**1668**:17

[196] Smith IM, McNamara DG, Scott PR, Holderness M. *Scirtothrips citri*. In: Quarantine Pests for Europe. 2nd ed. Wallingford: CABI/EPPO; 1997. p. 1425

[197] Pehrson JE, Flaherty DL, O'Connell NV, Phillips PA, Morse JG. Integrated Pest Management for Citrus. 2nd ed. Oakland, CA: University of California Statewide Integrated Pest

Management Project, DANR Publ. 3303; 1991

[198] Morse JG. Prospects for IPM of citrus thrips in California. In: Thrips Biology and Management. Boston, MA: Springer; 1995. pp. 371-379

[199] Perrotta G, Conti F. A threshold hypothesis for an integrated control of thrips infestation on citrus in south-eastern Sicily. IOBC/WPRS Bulletin. 2008;**38**:204-209

[200] Teksam I, Tunç I. An analysis of Thysanoptera associated with citrus flowers in Antalya, Turkey. Composition, distribution, abundance and pest status of species. Applied Entomology and Zoology. 2009;**44**:455-464

[201] Ferguson L, Grafton-Cardwell EE. Citrus Production Manual. Oakland, California: University of California, Agriculture and Natural Resources; 2014. p. 3539

[202] Vono G, Bonsignore CP, Marullo R. A comprehensive thrips species assessment for eco-consistent management of infestations in Mediterranean citrus crops. Horticulturae. 2022;**8**(2):137

[203] Arpaia ML, Morse JG. Citrus thrips *Scirtothrips citri* (Moulton) (Thys., Thripidae) scarring and navel orange fruit quality in California. Journal of Applied Entomology. 1991;**111**(1-5): 28-32

[204] Bravo-Mojica H. Ecological studies on the Citrus thrips, *Scirtothrips citri* (Moulton), (Thysanoptera, Thripiidae) in Southern California [Ph.D. dissertation]. Riverside: University of California; 1975. p. 159

[205] Tanigoshi LK, Nishio-Wong JY, Fargerlund J. Greenhouse-and laboratory-rearing studies of *Euseius hibisci* (chant) (Acarina: Phytoseiidae), a natural enemy of the citrus thrips, *Scirtothrips citri* (Moulton) (Thysanoptera: Thripidae). Environmental Entomology. 1983;**12**(4): 1298-1302

[206] Congdon BD, McMurtry JA. Biosystematics of Euseius on California citrus and avocado with the description of a new species (Acari: Phytoseiidae). International Journal of Acarology. 1985; **11**(1):23-30

[207] Morse JG, Zareh N. Pesticide-induced hormoligosis of citrus thrips (Thysanoptera: Thripidae) fecundity. Journal of Economic Entomology. 1991; **84**(4):1169-1174

[208] Grafton-Cardwell B, Eller A, O'Connell N. Integrated citrus thrips control reduces secondary pests. California Agriculture. 1995;**49**(2):23-28

[209] Badii MH, Hernández-Ortiz E, Flores AE, Landeros J. Prey stage preference and functional response of *Euseius hibisci* to *Tetranychus urticae* (Acari: Phytoseiidae, Tetranychidae). Experimental & Applied Acarology. 2004;**34**(3):263-273

[210] Grafton-Cardwell EE, Ouyang Y, Striggow RA. Predacious mites for control of citrus thrips, *Scirtothrips citri* (Thysanoptera: Thripidae) in nursery citrus. Biological Control. 1999;**14**(1): 29-36

[211] Navarro-Campos C, Beltrà A, Calabuig A, Garcia-Marí F, Wäckers FL, Pekas A. Augmentative releases of the soil predatory mite *Gaeolaelaps aculeifer* reduce fruit damage caused by an invasive thrips in Mediterranean citrus. Pest Management Science. 2020;**76**(7): 2500-2504

[212] Silva LP, Souza IL, Marucci RC, Guzman-Martinez M. *Doru luteipes* (Dermaptera: Forficulidae) and *Orius insidiosus* (Hemiptera: Anthocoridae) as nocturnal and diurnal predators of thrips. Neotropical Entomology. 2022; **1-10** (In Press)

[213] Yano E. Biological control using zoophytophagous bugs in Japan. Journal of Pest Science. 2022;**95**:1473-1484

[214] Baker G, Jackman DJ, Keller M, MacGregor A, Purvis S. Development of an integrated Pest management system for thrips in Citrus (Adelaide, SARDI). In: Horticulture Australia Final Report No CT97007. (Sustainable Systems) Adelaide: South Australia Research and Development Institute; 2000

[215] Hua LZ. List of Chinese Insects: Homoptera: Adelgoidea and Aphidoidea. Vol. 1. Guangzhou, China: Zhongshan University Press; 2000. pp. 110-131

[216] Khan I, Morse JG. Citrus thrips (Thysanoptera: Thripidae) resistance monitoring in California. Journal of Economic Entomology. 1998;**91**(2): 361-366

[217] Khan I, Morse JG. Impact of citrus thrips chemical treatments on the predatory mite *Euseius tularensis*. Journal of Applied Entomology. 2006;**130**(6-7): 386-392

[218] Ladaniya M. Citrus Fruit: Biology, Technology and Evaluation. Cambridge, Massachusetts, United States: Academic Press, Elsevier; 2010

[219] Prabhaker N, Castle SJ, Naranjo SE, Toscano NC, Morse JG. Compatibility of two systemic neonicotinoids, imidacloprid and thiamethoxam, with various natural enemies of agricultural pests. Journal of Economic Entomology. 2011;**104**(3):773-781

[220] Ujiye T. Biology and control of the citrus leafminer, *Phyllocnistis citrella* Stainton (Lepidoptera: Gracillariidae) in Japan. Japan Agricultural Research Quarterly. 2000;**34**(3):167-174

[221] Beattie GA. Integrated Control of Citrus Leafiminer. NSU Agriculture. Biological ana Chemical Research Institute; 1993

[222] Ware AB. The biology and control of citrus leafminer. Citrus Journal. 1994; **4**:26-28

[223] Heppner JB. Citrus leafminer, *Phyllocnistis citrella* in Florida (Lepidoptera: Gracillariidae; Phyllocnistinae). Tropical Lepidoptera Research. 1993;**04**:49-64

[224] Moy MA, Nguyen R. Classical biological control of the citrus leafminer: A progress report in Florida. Citrus Ind. 1994;**75**(6):61-62

[225] French JV, Legaspi JC. Citrus leafminer in Texas: Population dynamics, damage and control. In: Managing the Citrus Leafminer. Proceedings from an International Conference; Orlando, Florida. Gainesville: University of Florida; 1996. p. 80

[226] Johnson SJ, Vaughn A, Bourgeois WJ. Rearing and release methods for *Ageniaspis citricola* for a classical bio-control program of the citrus leafminer in Louisiana. In: Managing the Citrus Leafminer. Proceedings from an International Conference; Orlando, Florida. Gainesville: University of Florida; 1996. p. 63

[227] Malausa JC, Quilici S, Brun P. Status and First Studies on the Citrus Leafminer in France: Azur Coast. Corsica and Reunion Island; 1996

[228] Uygun N. Status of the citrus leafminer in Turkey. In: Proceeding of the International Meeting of Managing the Citrus Leafminer. Gainesville, Florida: University of Florida; 1996

[229] Quilici S, Franck A, Vincenot D, Montagneux B. Un nouveau ravageur des agrumes à la Réunion. La mineuse *Phyllocnistis citrella*. 1995

[230] Kerns D, Wright G, Loghry J. Citrus Leafminer (*Phyllocnistis Citrella*). Cooperative Extension. The University of Arizona; 2001

[231] Gabriel DW, Kingsley MT, Hunter JE, Gottwald T. Reinstatement of *Xanthomonas citri* (ex Hasse) and *X. phaseoli* (ex Smith) to species and reclassification of all *X. campestris pv. Citri* strains. International Journal of Systematic and Evolutionary Microbiology. 1989;**39**(1):14-22

[232] Stall RE, Seymour CP. Canker, a threat to citrus in the gulf-coast states. Plant Disease. 1983;**67**:581-585

[233] Schoulties CL, Civerolo EL, Miller JW, Stall RE, Krass CJ, Poe SR, et al. Citrus canker in Florida. Plant Disease. 1987;**71**(5):388-395

[234] Gottwald TR, Graham JH, Schubert TS. Citrus Canker: The Pathogen and its Impact. Online. Plant Health Progress; 2002

[235] Knapp JL, Browning HW, Albrigo LG, Pena JE, Stansly PA, Bullock RC. Management of the citrus leafminer: Chemical options. Citrus Industry (Bartow). 1996;**77**(3):48-49

[236] Pena JE, Duncan R. Control of the citrus leafminer in south Florida. Proceedings of the Florida State Horticultural Society. (Gainesville). 1993;**106**:47-51

[237] Hoy MA, Nguyen RU. Classical biological control of the citrus leafminer *Phyllocnistis citrella* Stainton (Lepidoptera: Gracillariidae): Theory, practice, art and science. Tropical Lepidoptera Research. 1997;**1997**:1-19

[238] Tan B, Huang M. Managing the citrus leafminer in China. In: Hoy MA, editor. Managing the Citrus Leafminer in Proceeding of the International Conference. Orlando, Florida, Gainesville: University of Florida; 1996. pp. 49-52. 119 pp

[239] Grafton-Cardwell EE. Kearney Agricultural Center, Parlier; Headrick DH, Cal Poly San Luis Obispo; K. E. Godfrey Calif. Dept. of Food and Agriculture; Kabashima JN. UC Cooperative Extension, Orange Co; Faber BA. UC Cooperative Extension, Ventura & Santa Barbara Co; Lazaneo VF. UC Cooperative Extension, San Diego Co; Mauk PA. Regional Director-Central Coast and Southern Region; Morse JG. Entomology, UC, 2021

[240] Browning HW, Pena JE, Stansly PA. Evaluating impact of indigenous parasitoids on populations of citrus leafminer. Folia Entomologica Mexicana. 1996;**79**:14-15

[241] PeñA JE, Duncan R, Browning H. Seasonal abundance of *Phyllocnistis citrella* (Lepidoptera: Gracillariidae) and its parasitoids South Florida citrus. Environmental Entomology. 1996;**25**(3): 698-702

[242] Browning HW. Biological control of the citrus leafminer by its native parasitoids and predators. Citrus Ind. 1995;**76**:46-48

[243] Pakistan Agricultural Research Council. Integrated pest management.

2006. Available from: http://www. parc. gov.pk/ipm.html

[244] Kunte K. Additions to known larval host plants of Indian butterflies. Journal of the Bombay Natural History Society. 2006;**103**:119-120

[245] Islam MS, Yasmin M, Islam AT. Studies on the biology and population abundance of lemon butterfly, *Papilio demoleus* L. (Papilionidae: Lepidoptera). Bangladesh Journal of Entomology. 2019; **29**:77-90

[246] Homziak NT, Homziak J. *Papilio demoleus* (Lepidoptera: Papilionidae): A new record for the United States, Commonwealth of Puerto Rico. Florida Entomologist. 2006;**89**(4): 485-488

[247] Patel PP, Patel SM, Pandya HV, Amlani MH. Survey on host plants and host plant preference by lemon butterfly *Papilio demoleus* Linnaeus (Lepidoptera: Papilionidae). Journal of Entomology and Zoology Studies. 2017;**5**(6):792-794

[248] Guerrero KA, Veloz D, Boyce SL, Farrell BD. First New World documentation of an Old World citrus pest, the lime swallowtail *Papilio demoleus* (Lepidoptera: Papilionidae), in the Dominican Republic (Hispaniola). American Entomologist. 2004;**50**(4): 227-229

[249] Larsen TB. Extension recente en Iraq de l'aire de *Papilio demoleus*. Linn: Entomops. 1977;(42):37-38

[250] Badawi A. Studies on some aspects of the biology and ecology of the citrus butterfly *Papilio demoleus* L. in Saudi Arabia (Papilionidae: Lepidoptera). Zeitschrift fuer Angewandte Entomologie. 1981;**91**(3):286-292

[251] Farid A. Some bio-ecological features of citrus butterfly in south-East Iran. Entomologie et Phytopathologie Appliquees. 1987;**54**(1-2):129-137

[252] Moonen JJM. *Papilio demoleus* L. in Java (Lep: Papilionidae). Transactions of the Lepidopterological Society of Japan. 1991;**42**(2):93-94

[253] Dunn KL. Butterfly observations in Southeast Asia including two potentially new records for Borneo. Victorian Entomologist. 1999;**29**(6):114-116

[254] Matsumoto K. *Papilio demoleus* (Papilionidae) in Borneo and Bali. The Journal of the Lepidopterists' Society. 2002;**56**:108-111

[255] Moonen JJM. *Papilio demoleus* L. (Lepidoptera: Papilionidae) in West Irian. Transactions of the Lepidopterological Society of Japan. 1999;**50**(2):82-84

[256] Smithers CN. Migration records in Australia. 2. Hesperiidae and Papilionidae (Lepidoptera). Australian Entomological Magazine. 1978;**5**(1): 11-14

[257] Williams AE, Scanlon MD, Himbeck KJ. New records of butterflies (Lepidoptera) from Dorre Island, Western Australia. Victorian Entomologist. 1998;**28**(3):55-58

[258] Xingyong C, Ruizhen W, Haidong C. Observation on the bionomics of *Papilio demoleus* in Guangzhou. Kun Chong Zhi Shi= Kunchong Zhishi. 2004;**41**(2):169-171

[259] Jahnavi M, Ramakrishna RA, Sarada G. Biology and morphology of citrus butterfly *Papilio demoleus* Linnaeus (Lepidoptera: Papilionidae) on acid lime. Journal of Entomology and Zoology Studies. 2018;**6**(1): 1556-1561

[260] Dharavath S, Karabhantanal S, Anand V, Akash MP. Biology of citrus butterfly *Papilio demoleus* L. (Papilionidae: Lepidoptera). Journal of Entomological and Zoology Studies. 2021;**9**:235-239

[261] Chatterjee H, Ghosh J, Senapati SK. Influence of important weather parameters on population fluctuation on major insect pest of mandarin orange (*Citrus reticulata* Blanco) at Darjeeling district of West Bengal (India). Journal of Entomological Research. 2000;**24**(3): 229-233

[262] Larsen TB. The zoogeographical composition and distribution of the Arabian butterflies (Lepidoptera; Rhopalocera). Journal of Biogeography. 1984;**11**:119-158

[263] Narayanamma VL, Savithri P, Rao AR. Influence of citrus butterfly *Papilio demoleus* L. damage on growth parameters of the sweet orange host plant. Indian Journal of Plant Protection. 2001;**29**(1-2):140-141

[264] Rafi MA, Matin MA, Khan MR. Number of generations and their duration of lemon butterfly (*Papilio demoleus* L.) in rainfed ecology of Pakistan. Pakistan Journal of Scientific Research. 1999;**51**(3-4):131-136

[265] Winotai A, Napompeth B. Natural enemies of the lemon butterfly, *Papilio demoleus* L.(Lepidoptera: Papilionidae) in Thailand. In: Natural Conference on the Progress of Biological Control in Thailand; 22-24 Jun 1981; Bangkok (Thailand).

[266] Lewis DS. Lime swallowtail *Papilio demoleus* Linnaeus (Insecta: Lepidoptera: Papilionidae). IFAS Extension University of Florida; 2018. pp. 1-5. Available from: https://citeseerx.ist.psu.edu/viewdoc/download?doi=10.1.1.739.130&rep=rep1&type=pdf

[267] Narayannamma VL, Savithri P, Rao AR. Natural enemies and per cent parasitisation of citrus butterfly, *Papilio demoleus* on sweet orange and acid lime. Indian Journal of Entomology (India). 2003;**65**:461-464

[268] Offor C, Okaka A, Aja P. Comparative analyses on the effects of ethanol leaf-extracts of *Vernonia amygdalina* and *Azadirachta indica* on lipid profile in Albino Rats. World Applied Sciences Journal. 2014;**32**:61-65

[269] Vattikonda SR, Sangam SR. Assessment of antifeedant potential of azadirachtin, a plant product of *Azadirachta indica* against *Papilio demoleus* L.(Lepidoptera: Papilionidae) larvae. Journal of Entomology and Zoology Studies. 2016;**4**(5):956-959

[270] Yasui H, Kato A, Yazawa M. Antifeedants to armyworm, *Spodoptera litura* and *Pseudaletia separata*, from bitter gourd leaves, *Momordica charantia*. Journal of Chemical Ecology. 1998;**24**:803-813

[271] Vattikonda SR, Amanchi NR, Raja S. Antifeedant activity of forskolin, an extract of coleus forskohlii, against *Papilio demoleus* L. (Lepidoptera: Papilionidae) larvae. European Journal of Experimental Biology. 2014;**4**(1): 237-241

[272] Vattikonda SR. Effect of andrographolide on feeding behaviour of *Papilio demoleus* L. (Lepidoptera: Papilionidae) larvae. Asian journal of Bioscience. 2015;**10**(1):65-70

[273] Moorthi V, Balasubramanian PC, Avery PB, Selvarani S, Kubendran T, Rathinakumar T, et al. Pathogenicity and proteome production of *Isaria*

fumosorosea (= *Paecilomyces fumosoroseus*) (WISE) isolates against lemon butterfly, *Papilio demoleus* (Papilionidae: Lepidoptera). African Journal of Biotechnology. 2014;**13**(43)

[274] Khan MA, Paul B, Ahmad W, Paul S, Aggarwal C, Khan Z, et al. Potential of *Bacillus thuringiensis* in the management of pernicious lepidopteran pests. In: Hakeem KR, Akhtar MS, editors. Plant, Soil and Microbes: Mechanisms and Molecular Interactions. Vol. 2. Cham, Switzerland: Springer International Publishing; 2016. pp. 277-301

[275] Osouli S, Afsharmanesh H. To evaluate the effects of secondary metabolites produced by *Bacillus subtilismutant* M419 against *Papilio demoleus* L. and *Aspergillus flavus*. Acta Ecologica Sinica. 2016;**36**(6): 492-496

[276] Hussain SI, Zahid M. Comparative effect of Biosal® and pyrethroids (deltamethrin and lambda cyhalothrin) on enzymatic activity and total protein contents in *Hermolaus modestus*. FUUAST Journal of Biology. 2016;**6**(2): 241-246

[277] Nikhat Y, Naqvi SNH, Khan MF, Arshad M, Tariq RM. Role of neem product (Biosal®) and its impact on cholinesterase and protein contents in larvae of *Papilio demoleus* L. in comparison with cypermethrin. Journal of Experimental Zoology. 2010;**13**(2): 541-544

[278] Haque R, Maleque MA, Rahman SM, Khan AU, Bhuiyan ML. Evaluation of new management approaches against lemon butterfly (*Papilio demoleus* l.) infesting jara lemon in Sylhet, Bangladesh. Bangladesh Journal of Entomology. 2019; **29**(2):1-2

[279] Singh S, Kumar K. Diofenolan: A novel insect growth regulator in common citrus butterfly, *Papilio demoleus*. Phytoparasitica. 2011;**39**(3): 205-213

[280] Somasundaram P, Chockalingam S. Impact of chitin synthesis inhibitor diflubenzuron on the feeding physiology of *Papilio demoleus* (Lepidoptera: Papilionidae). International Journal of Tropical Insect Science. 1988;**9**(5): 613-616

[281] Thakur K, Sharma A, Sharma K. Management of agricultural insect pests with physical control methods. The Pharma Innovation Journal. 2021;**10**: 306-314

Section 3

Citrus in Human Health

Chapter 9

Citrus: An Overview of Food Uses and Health Benefits

Sakhawat Riaz, Arslan Ahmad, Rimsha Farooq, Nasir Hussain, Tariq Riaz, Khadim Hussain and Muhammad Mazahir

Abstract

Citrus species is a category of fruit that contains a variety of bioactive components throughout the plant. *Citrus* fruits (and items made from them) are among the most widely eaten fruits in the world, and their supply continues to increase. Oranges, pomelos, limes, tangelos, mandarins, lemons, kumquats, grapefruits, and other *Citrus* fruits are among them. They are frequently employed in the culinary, cosmetics, and pharmaceutical sectors due to their fragrance and taste. Vitamin C, pectin, limonene, phenolics, iso-limonene, flavanones, and nonanal are the main bioactive components present, and they provide a variety of health advantages. Pharmacological studies have shown that the fruit has numerous nutraceutical benefits, including a strong antioxidant, antidiabetic, anti-hypertensive, anticancerous, antibacterial, antifungal, antimicrobial, antihyperglycemic, and cardioprotective. It should also be highlighted that all *Citrus* fruits are an excellent source of minerals, which are required to maintain water and electrolyte balance. *Citrus* fruit-eating has been linked to a range of health advantages in recent research. This chapter presents an overview of the nutritional aspects of *Citrus* as well as its health benefits, which will be detailed.

Keywords: *Citrus*, nutrition, bioactive substance, healthy food, vitamin, antioxidant, phenolic compound

1. Introduction

Citrus fruits are members of the *Rutaceae* family and are the most marketed natural product on earth. *Citrus*, also known by the Romance loanword agrumes (sour fruits), is one of the world's dominant cultivars, with a feature of the application and acceptance significantly to human food [1]. *Citrus,* which is commonly grown in subtropical and tropical regions, is one of the largest and most significant crops. Annually, over 150 million tons of *Citrus* natural products are filled in more than 140 nations. According to Food and Agriculture Organization (FAO), Italy, Spain, Argentina, the United States, and Greece are significant exporters of fresh *Citrus*, while the main exporters of organic *Citrus* juices are Brazil, Israel, Costa Rica, the United States, Italy, Mexico, and Cuba [2]. *Citrus* natural products have been appreciated by people since ancient times. *Citrus* fruits are an important part of today's diet since they are soft, easy to peel, and have juicy flesh with a distinct flavor. Many

Citrus species, such as grapefruits (*Citrus paradisi*), mandarins (*Citrus reticulata)*, lemons (*Citrus limon*), oranges (*Citrus sinensis)*, *Citrus clementina*, and *Citrus unshiu* are consumed as juice or used for the development of new products all over the planet [3]. *Citrus* is popular due to its distinct flavor, taste, and scent, as well as its increased level of phenolics, vitamin C, and other nutritional properties [4]. *Citrus* belongs to the *Citrinae* subtribe, tribe *Citreae*, and subfamily *Aurantioideae*. Furthermore, the continuing ordered study seems, by all accounts, to be confounded and disputed, as a result of sexual suitability among *Citrus* species and other taxa, a frequency range of bud modifications, and apomixis (e.g., adventitious embryonic) [5]. Swingle and Reece and Tanaka's taxonomy classifications for *Citrus,* which classified 16 and 162 species, respectively, are the most frequently accepted [6]. Fruit and vegetable-rich cuisines have been significantly linked to multiple therapeutic benefits and a decreased probability of having illness [7]. *Citrus* is one of the world's most significant natural product crops, grown for both fresh juice production and food processing. *Citrus* fruits and commodities are quite common with significant economic and nutritional influence in both developing and developed nations due to inexpensive economic accessibility, and customer attitude toward the increasingly recognized potential health benefits [8].

Dietary choices have been related to a range of health outcomes, including high blood cholesterol, hyperglycemia, high blood pressure, and other major chronic degenerative conditions [9]. There is a huge necessity for nutritious meals that, in addition to providing nourishment, may improve the public overall health status. Fruits and vegetables are high in nutrients, anti-inflammatory compounds, and phytochemicals [10]. Aside from that, the demand for plant-based commodities is increasing in both emerging and developed countries [11]. Thus, we may employ specific fruits, such as citron (*C. medica*), in various formulations and benefit from their health capabilities in lowering the risk of numerous prevalent ailments. The World Health Organization (WHO) suggests, including 400 g of fruits, in your balanced meals. Consuming more *Citrus* fruits, vegetables, and whole grains has been associated with a decreased risk of numerous metabolic syndrome-related disorders, particularly malignancy, impaired glucose tolerance, coronary heart disease, and neurodegenerative diseases, according to deterministic, observational, and intervention studies [12]. These illnesses are essentially connected with serious and second-rate extensive irritation created by responsive oxygen species. The bioactive constituents included in whole grains, vegetables, and fruits protect cells against oxidative damage by neutralizing free radicals, hence lowering the prevalence of these disorders [13]. *Citrus* fruit species have been extensively researched for their bioactive content and medicinal properties. Oranges account for the majority of all *Citrus* production/exports, followed by lemons/limes (8%), grapefruit (7%), and tangerines/mandarins (one-third) [14]. *Citrus* fruits have gained popularity in recent years due to their health-promoting properties. A variety of active substances contained in *Citrus* natural products have been detached and portrayed, and their part in people's well-being has been extensively researched. *Citrus* fruits' significance in one's nutrition, as well as the abundance of key bioactive compounds, such as ascorbic acid (AsA), carotenoids, and flavonoids, in *Citrus* fruits, and their therapeutic properties are discussed in this chapter.

2. Nutritional perspectives of *Citrus*

Citrus spp. is abundant in vitamin C and is low in cholesterol, salt, and fat, and has a relatively low average calorie value, which may be helpful for users worrying about

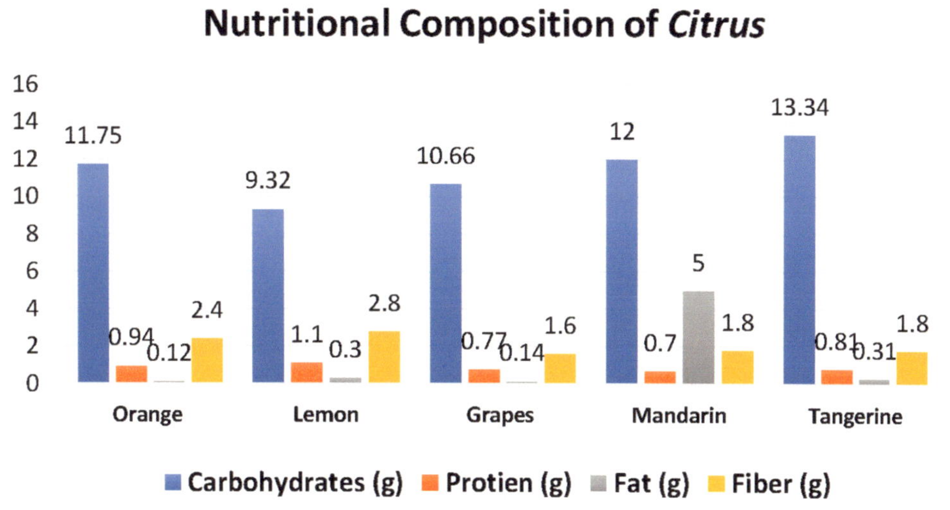

Figure 1.
Nutritional composition of Citrus *fruit [1].*

being overweight. *Citrus* includes significant levels of carotenoids (some of which can be converted to vitamin A), folate, and fiber [15]. The macronutrient composition of some *Citrus* fruit is presented in **Figure 1** [1].

The fruits are high in important nutrients, such as dietary fiber, and simple sugars, and often a variety of micronutrients, including thiamin, copper, riboflavin, vitamin B6, magnesium, niacin, calcium, potassium, folate, phosphorus, pantothenic acid, all of which have been required for health maintenance and proper development [16]. Additionally, studies were carried out to better comprehend the diversity found in nature phytonutrients, such as flavonoids, carotenoids, and limonoids. Epidemiological data and other investigations have revealed that all these active components have a broad range of biological consequences and they may serve to act as an intermediary between *Citrus* fruit consumption and sickness protection [17], such as hyperglycemia, cancer, osteoporosis, cataracts, and cardiovascular disease. The nutritional profile of various *Citrus* fruits is presented in **Table 1** [1].

3. Bioactive compounds of *Citrus* fruits

Citrus fruits are high in bioactive chemicals, particularly phenolic compounds (flavonoids, coumarins, and phenolic acids), terpenoids (carotenoids and limonoids), and pectin [3, 18, 19]. *Citrus* fruits are also high in nutrients, including ascorbic acid, tocotrienols, tocopherols, and minerals (iron, manganese, zinc, selenium, and copper) [3, 18, 19]. Flavonoids, which are a significant source of antioxidants in the diet of humans, are a kind of polyphenolic compound abundant in *Citrus* fruits. Flavanones account for approximately 95% of the total flavonoids generated by *Citrus* [20]. *Citrus*-derived flavonoids can be aglycones or glycosides, and they as a rule do not happen in nature as aglycones yet rather as glycosides, in which the aglycones are linked to a sugar molecule [21]. *Citrus* fruits contain a significant quantity of flavonols (kaempferol, quercetin, and rutin), polymethoxylated flavones (e.g., 5-dimethyl nobiletin, tangeritin, and nobiletin,), flavanone-7-O-glycosides (narirutin, naringin,

Nutritional characteristics	Lemon	Grapefruit	Orange	Tangerine
Energy (kcal)	29	42	47	53
Dietary fiber (g)	2.80	1.60	2.40	1.80
Protein (g)	1.10	0.77	0.94	0.81
Carbohydrates (g)	9.32	10.66	11.75	13.34
Cholesterol (g)	0	0	0	0
Total fat (g)	0.30	0.14	0.12	0.31
Niacin (mg)	0.100	0.204	0.282	0.376
Riboflavin (mg)	0.020	0.031	0.040	0.036
Pantothenic acid (mg)	0.190	0.262	0.250	0.216
Pyridoxine (mg)	0.080	0.053	0.060	0.078
Vitamin A (IU)	22	1150	225	681
Vitamin C (mg)	53	31.20	53.20	26.70
Vitamin E (mg)	0.15	0.13	0.18	0.20
Calcium (mg)	26	22	40	37
Potassium (mg)	138	135	181	166
Copper (μg)	37	32	45	42
Magnesium (mg)	8	9	10	12
Iron (mg)	0.60	0.08	0.10	0.15
Zinc (mg)	0.06	0.07	0.07	0.07
Manganese (mg)	0.030	0.022	0.025	0.039
α-Carotene (μg)	1	3	11	101
β-Carotene (μg)	3	686	71	155
Xanthophylls (μg)	11	5	129	138
β-Cryptoxanthin (μg)	20	6	116	407
Lycopene (μg)	0	1419	0	0

Table 1.
Nutritional characteristics for Citrus *fruits per 100 g fruit [1].*

eriocitrin, and hesperidin), flavones (e.g., vitexin, diosmin, and rhoifolin,), and anthocyanin (peonidin glucosides and cyanidin) [22, 23]. When contrast to other *Citrus* species, *C. aurantium* has a higher concentration of active alkaloids, particularly synephrine, which accounts for more than 85% of the total alkaloid value [24]. The most common limonoids due to the dominance of *Citrus* species are limonin and limonin glucoside. Carotenoids are a type of isoprenoid pigment that is widely used in photosynthesis and signaling [24]. *Citrus* fruits' peel and pulp are orange-red owing to the existence of carotenoids and apocarotenoids [25]. *Citrus* fruits' carotenoid content is by carotenoid fatty acid esters (xanthophyll esters) [25, 26]. The bioactive compound of *Citrus* is presented in **Table 2** [1].

The existence of specific total carotenoids & xanthophyll esters is strongly impacted by species, maturation stage, and fruit components. Tangerines and oranges contain substantial amounts of β-cryptoxanthin, zeaxanthin Lutein, and β-carotene [27].

Fruit	Metabolites	Flavonoid	(mg/100 g)
Grapefruit	Flavanones	Naringenin	21.34
		Eriodictyol	0.65
		Hesperetin	2.35
	Flavonols	Quercetin	0.40
		Myricetin	0.05
Grapefruit, pink and red	Flavones	Luteolin	0.60
	Flavonols	Quercetin	0.33
		Kaempferol	0.01
		Myricetin	0.01
	Flavanones	Naringenin	32.64
		Hesperetin	0.35
Grapefruit white	Flavanones	Naringenin	31.18
Lemons	Flavonols	Quercetin	1.14
		Kaempferol	0.03
		Myricetin	0.50
	Flavanones	Naringenin	0.55
		Eriodictyol	21.36
		Naringenin	27.90
	Flavones	Luteolin	1.90
Lime	Flavonols	Quercetin	0.40
	Flavanones	Naringenin	3.40
		Hesperetin	43.00
Orange	Flavonols	Quercetin	0.45
		Kaempferol	0.13
		Myricetin	0.15
	Flavanones	Naringenin	15.32
		Hesperetin	27.25
		Eriodictyol	0.17
Tangerine	Flavanones	Naringenin	10.02
		Hesperetin	7.94

Table 2.
Bioactive substance of Citrus *fruits [1].*

The carotenoids found in the endocarp and flavedo of completely developed oranges were (all-E) and (9Z) violaxanthin, which were esters and monoesters containing myristate, palmitoleate, caprate, stearate, palmitate, oleate acyl moieties & laurate, lutein, β-carotene, and antheraxanthin were the other important carotenoids. In contrast, violaxanthin, β-carotene, α-carotene, and lutein were shown to be plentiful in ripe green fruit, furthermore, β-citraurin esters were discovered in *Citrus* fruit flavedo [28]. *Citrus* fruits' provitamin a carotenoids (e.g., β-cryptoxanthin) have also been found to help with metabolic disorders, such as type 2 diabetes [29]. Hesperidin (HSP) is a potent bioactive flavonoid aglycone and subclass of flavonoids found in *Citrus* species (lime, orange, lemon, and blood orange) [30]. This flavanone has been demonstrated to have a variety of pharmacological actions, including antiviral activities, anticarcinogenic, anti-inflammatory, and antioxidant properties, analgesic [31], hypolipidemic, hypoglycemic activities, and anticoagulant [32].

4. Health perspectives

Citrus fruit has natural compounds, which offer a wide range of bioactivities, including antimicrobial, anti-sensitivity, antioxidant, and anti-disease capabilities, as well as heftiness, hepatoprotective, neuroprotective, and cardiovascular regulating qualities. The different medicinal characteristics of *Citrus* are layout in **Figure 2**.

4.1 *Citrus* fruits and (COVID-19)

The potential significance of diet in the mitigation of coronavirus infection 2019 (COVID-19) has been undervalued so far among the different options. Foods contain usually compounds that may have a health benefit, and some of these chemicals may have antiviral properties and may be useful in altering the immune function and source of antioxidants from the oxidative stress generated by infection [37]. *Citrus* fruits are high in flavanones, vitamin C, and anthocyanins the most abundant of which are naringin and hesperidin, both of which have antioxidant and anti-inflammatory qualities [38–40]. Fibers, such as pectin, which is more abundant in the solid section, aid in the regulation of intestinal activities and the prevention of LDL cholesterol absorption. *Citrus* may also be effective in reducing the risk and cure of viral diseases [37].

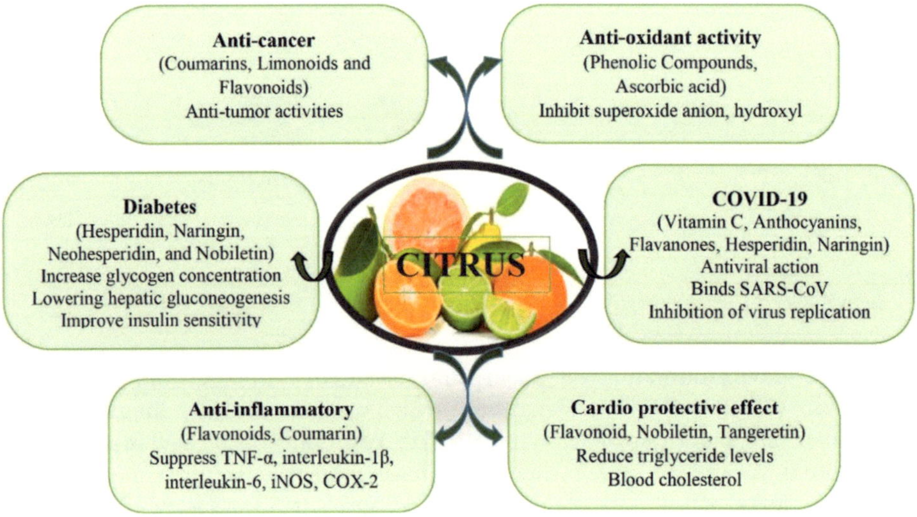

Figure 2.
Health benefits of Citrus *and its constituent [27, 33–36].*

The revelation that hesperidin possesses a chemical-physical structure that allows it to attach to critical proteins involved in the SARS-CoV-2 virus's activity has piqued scientists' curiosity. At least six investigations returned similar outcomes [41–45].

Wu and collaborators [41] for binding to SARS-CoV-2 proteins, evaluated 1066 natural compounds with the ability of antiviral effects, along with 78 antiviral medicines that have previously been described. The best option for bonding to the "spike" was hesperidin. At the point when the ACE2 — RBD complex is superimposed over the hesperidin — RBD complex, there is critical hesperidin contact with the ACE2 interface, showing that hesperidin may alter ACE2's interaction with RBD. Another thorough molecular docking investigation of the hesperidin-Mpro interaction was published recently [44]. The lowest binding energy (showing the most affinity) was revealed to be rutin (9.55 kcal/mol), followed by hesperidin (9.02 kcal/mol), emetine (9.07 kcal/mol), ritonavir (9.52 kcal/mol), and indinavir (8.84 kcal/mol) in a screening of 33 natural and already known antiviral compounds. THR45, THR25, THR24, CYS145, HIS4, and SER46 are among the amino acids to which hesperidin forms hydrogen bonds. Joshi et al. research provided more evidence [45], who discovered hesperidin as one of the numerous natural compounds that adhere to the primary protease of the SARS-CoV-2 virus. In addition, the viral receptor angiotensin-converting enzyme-2 is involved (ACE-2).

As a powerful antioxidant against superoxide and hydroxyl radicals, hesperidin plays an important role in antioxidant defense mechanisms [46], and hesperetin, a derivative of it, prevents LPS-stimulated microglial cells from producing nitric oxide [47]. Another research found that *Citrus* flavanones, such as naringenin and hesperidin, reversed age-related decreases in superoxide dismutase, catalase, and glutathione reductase in the livers of elderly rats [48]. Coronaviruses are one type of virus that causes the common cold, a condition for which there is no treatment or vaccination. Given the fact that SARS-CoV is a coronavirus, as well as the low cost and high security of fresh foods with high doses of vitamin C, it has been proposed that boosting regular consumption of these foods may be advantageous during the COVID-19 pandemic [49–51]. **Figure 3** shows hesperidin and L-ascorbic acid reduces the cell pattern of the SARS COVID-2 (SARS-CoV-2) infection, as well as the regions of infection, started cell, and fundamental illness (set apart with an "X") [37].

As well as aiding the development of collagen in connective tissue, L-ascorbic acid has antioxidant capacity, and when combined with other minerals, enzymes, and vitamins, can lessen the effects of free radicals. Human vascular smooth muscle cells are considered to be protected by vitamin C from apoptosis by preventing LDL oxidation [52]. *Citrus* fruits may have a substantial impact on COVID-19 therapy, through pathways other than viral replication suppression and antioxidant activity [37]. The numerous biological effects of vitamin C and hesperidin are two core aspects of *Citrus* fruits that appear to be excellent choices for counteracting SARS-CoV-2 cell infiltration and modulating the disease's systemic immuno-pathological stages. The more epidemiological, preclinical, and clinical study is required to verify the concept that a diet rich in *Citrus* fruits or comparable compounds may help COVID-19 prevention efforts [37].

4.2 Antioxidant activity

The primary source of reactive oxygen species (ROS) is oxygen, in living organisms through a variety of metabolic routes, such as hydroxyl radicals, hydrogen

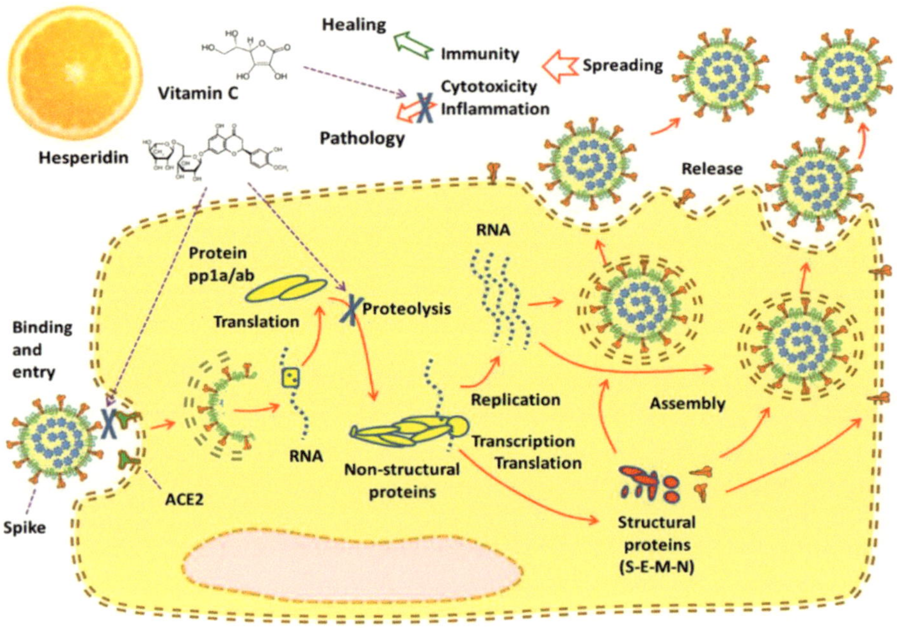

Figure 3.
Hesperidin and vitamin C impact on infected cell cycle [37].

peroxide, and superoxide anion, while antioxidant systems may combat them to maintain equilibrium [53]. However, contemporary lifestyle variables may increase the number of reactive oxygen species, which play an important role in the development of different diseases, such as inflammation, heart disease, arthritis, aging, and cancer, and produce oxidative stress. *Citrus* extracts, including *Citrus* bergamia juice extracts, karna peel extracts, and *Citrus limetta* peel extracts, have been demonstrated to have antioxidant activity [33, 34]. *Citrus* fruits are said to be high in antioxidants because they contain phenolic compounds with poly-hydroxyl groups, such as flavonoids, phenolic acids, and their derivatives [54]. The following are the major antioxidant mechanisms:

- Direct absorption and neutralization of free radicals.
- Suppression of ROS-related enzymes: xanthine oxidase, myeloperoxidase, and NADPH oxidase [55].
- Increased activity of mammalian antioxidant enzymes, such as superoxide dismutase, catalase, and others [56].

4.3 Hepatic protective effects of *Citrus* fruit

The liver, the body's most significant digestive organ, is responsible for toxic compound metabolization via several routes, such as hydrolysis, oxidation, reduction, conjugation, and hydration [57]. Lipopolysaccharide (LPS) is a glycolipid of gram-negative bacteria's cell wall [16]. LPS raises total bilirubin levels, alkaline phosphatase (ALP), aspartate aminotransferase (AST), and alanine aminotransferase (ALT). Tissue and serum nitrite levels, as well as thiobarbituric acid reactive substances (TBARS) levels, increased, whereas superoxide dismutase (SOD) and glutathione (GSH) content decreased. However, hesperidin injection restored all of these abnormalities, and the authors concluded that hesperidin might reduce nitric oxide and reactive oxygen species (ROS) formation and prevent LPS-induced hepatotoxicity [17].

4.4 Anti-inflammatory

Inflammation is a complicated process characterized primarily by inflammatory cytokines, such as interleukin-6, tumor necrosis factor-alpha (TNF-α), interleukin-1β, and also a chain of molecular messengers, such as inducible nitric oxide synthase (iNOS) and cyclooxygenase-2 (COX-2). These inflammatory cytokines have also been linked to the development of several inflammatory disorders, particularly colorectal cancer, Alzheimer's disease, multiple sclerosis, and Parkinson's disease [35]. Orange (*Citrus aurantium L.)* isolate has been shown to inhibit UVB-induced COX-2 expression and prostaglandin (PG) E2 generation in HaCaT cells while also acting as a PPAR-c agonist [58]. *Citrus* fruit contains volatile oil coumarin and flavonoids, which have anti-inflammatory properties and can be utilized as a remedy to prevent or treat chronic inflammatory illnesses.

A meal incorporating 0.1% rutin, in particular, alleviated dextran sulfate sodium (DSS)-induced colon inflammation in mice, increasing histological ratings and mitigating weight loss, presumably via lowering pro-inflammatory cytokine production [59]. Furthermore, studies suggest that quercetin (50 and 100 mg/kg) treatment lowered structural, clinical, and behavioral outcomes in rats' colons [60]. Isoquercitrin, a flavonoid with anti-inflammatory properties and antioxidants, is a glycoside that is chemically related to quercetin. This flavonoid reduces the levels of cyclo-oxygenase-2 (COX-2) and iNOS in rats, which helps them recover from acute colitis [61]. Kumar et al. observed that administering naringin at various dosages decreased acetic acid-induced colitis in animal models by decreasing DNA damage and inflammatory responses [62]. Naringenin is abundant in *Citrus* and has been extensively researched for its potential efficacy in several animal models of inflammatory illness. Apigenin is a flavone that may be found in a variety of fruits and vegetables, with grapefruit having the highest concentration. Its potential to relieve symptoms of inflammatory diseases is considered to be due to both anti-inflammatory and antioxidant properties. This was achieved by activating the aryl hydrocarbon (Ah)-receptor, which in turn produced good protective enzymes and cytokines, resulting in an improvement in the anti-inflammatory system [63]. Diosmin, a *Citrus* flavone recognized for its anti-inflammatory, antioxidant, and vasotonic effects, has also been researched for its potential to attenuate experimentally-induced colitis in animal models [64]. Among flavones, luteolin has proven tremendous anti-inflammatory efficacy in lots of experimental models. Luteolin inhibited intracellular inflammatory send signal in HT-29 colon epithelial by downregulating the Janus kinase (JAK)/sign transducer and activator of transcription (STAT) pathway [65]. Nobiletin and tangeretin are polymethoxylated flavones determined in *Citrus,* which have been established to assist in inflammation in many studies. In each *in vitro* and *in vivo* experiments, nobiletin suppressed inflammatory processes, along with the suppression of iNOS and COX-2 expression, and repaired the impaired intestinal barrier characteristic in trinitrobenzene sulfonic acid

(TNBS)-induced colitis in rats and Caco-2 monolayer inhibiting the protein kinase b (Akt)-NF-jB-myosin light-chain kinase (MLCK) pathway affects the latter [66].

4.5 Anti-cancer

Secondary metabolites in *Citrus* fruits, such as coumarins, limonoids, and flavonoids, have been linked to a lower risk of cancer, including lung tumorigenesis, colonic tumorigenesis, breast cancer, hematological malignancies, gastric cancer, and hepatocarcinogenesis, among others [66–70]. Chang and Jia observed that the flavedo fraction of organ (*Citrus reticulata cv. Suavissima*) inhibited tumor growth by blocking epithelial-to-mesenchymal transformation and interfering with the (transforming growth factor-β1) TGF-β1–SMAD (Suppressor of Mothers against Decapentaplegic) Snail/Slug axis [71]. *Citrus* may lower cancer by inhibiting oxidative stress and damage, as well as interfering with cancer initiation, development, and progression [72]. *Citrus bergamia* (bergamot) juice has subsequently been revealed to exhibit anti-cancer potential in several *in vitro* and *in vivo* studies, with the flavonoid content attributed to this effect [73]. Vitamin C has been suggested to combat inflammation and consequent oxidative damage to DNA, both of which have a role in the beginning and progression of cancer. Furthermore, because of its pro-oxidant activity, vitamin C can destroy cancer cells [36].

4.6 Cardiovascular protection properties

Several recent epidemiologic studies commonly correlate greater flavonoid-rich food consumption to lower cardiovascular morbidity and mortality [74]. Several studies have demonstrated that *Citrus*-derived flavonoids may reduce triglyceride levels (TG) and blood cholesterol (CH). Using HepG2 cells, the optimum structure was found to be complete methoxylation of the A-ring of *Citrus* flavonoids for expressing a profound impact on regulating hepatic metabolic activity by decreasing apoB-containing lipoprotein production [75]. Nobiletin and tangeretin, which have the ideal chemical composition, may reduce blood triglycerides, but several *Citrus* flavonoids, such as naringin and hesperidin, that lack a methoxylated A-ring might have little or no lipid-lowering activity *in vivo* [76].

4.7 Impact on hyperglycemia

Citrus flavonoids (naringin, nobiletin hesperidin, and neohesperidin) reduced amylase-catalyzed starch digestion considerably. Furthermore, neohesperidin and naringin hindered only amylose digestion, but nobiletin and hesperidin inhibited both amylopectin and amylose metabolism. This research shows that *Citrus* flavonoids had an essential part in avoiding hyperglycemia development, in part by interacting with starch, improving hepatic metabolism and glycogen content while inhibiting hepatic gluconeogenesis [77]. Naringin, hesperidin, and nobiletin were also shown to have anti-diabetic properties, perhaps through increasing insulin sensitivity or increasing hepatic gluconeogenesis in hyperglycemic mice [78]. According to one research, streptozotocin nicotinamide-induced experimental in diabetic rats, naringenin protects against diabetes by exerting antihyperglycemic and antioxidant properties [79]. Chronic naringenin therapy in diabetic animals might avoid functional alterations in vascular reactivity via a prostaglandin-independent and NO-dependent mechanism [80].

5. *Citrus* byproducts utilization

Citrus juice, the principal product of processing firms is strong in vitamin C and is frequently employed in the production of nutrient-dense drinks. A large portion of this garbage is thrown on nearby landfills or burnt, resulting in contamination and a decrease in the dissolved oxygen concentration of contaminated water. The best way to handle these residues is to extract macro and micronutrients from by-products, use fiber-rich components in confectionery goods, fortify nutrient-rich animal feed, and bio-oils, and produce organic fertilizers, ethanol, and essential oils [81].

Citrus peel wastes are partially utilized in livestock feed, either fresh or after dehydration, but a significant amount of *Citrus* pulp is ended up lost in the fresh state owing to the complexity of handling and disposal of a significant amount of garbage generated in a very short period, and a considerable portion of it is nevertheless discarded into the atmosphere, possibly creating several ecological difficulties. As a result, the disposal of these *Citrus* by-products is a major issue for the *Citrus* industry all over the world [82].

Many studies are available on the handling of *Citrus* trash for the extraction of organic value-added substances, such as fiber and bioactive chemicals, such as flavonoids [83]. After processing *Citrus* fruits, more than 60 thousand tons of *Citrus* pomaces (CPs) are generated in South Korea each year. Public demand for non-synthetic, better natural food raw materials has fueled research into the recuperation of natural value-added chemicals from *Citrus* trash [84].

Citrus waste phytochemicals and value-added substances are used in the development of healthful meals, flavoring enhancers in food processing, health and power beverages, preservatives, and vitamin supplements. These aid in improving the flavor and scent of meals as well as correcting inadequacies. *Citrus* waste-derived phytochemicals are also used in skin, hair, and nail care products, as well as antibacterial antifungal lotions, toiletries, fragrances, and soaps [85].

Mucilage & pectin are elevated chemicals generated from *Citrus* waste. Pectin is a natural vegetable substance that is useful as a cosmetic and nutritional supplement, in pharmaceutical sectors owing to its stabilizing, thickening, and gelling qualities [86]. They are starches classified as "dietary fiber." Mucilages are the soluble dietary fiber that may be discovered in *Citrus* waste as well as other plants. They are also vegetable polysaccharides that are identical to pectin but varied in their uronic acid and sugar content. They, like pectin, may be employed in the culinary, nutritional, cosmetic, and pharmaceutical industries; new research has emphasized the anti-inflammatory properties of lemon mucilage [87].

Flavonoids, which are known antioxidants and are abundant in fruits and vegetables, are another fascinating substance that may be derived from *Citrus* trash. The same is used in both the pharmaceutical and food industries. Hesperidin, which can be isolated from orange and lemon peel, is particularly useful in the pharma industry due to its anti-inflammatory and vasodilatory effects [87].

Citrus trash may be converted into limonene, ethanol, and other byproducts. Limonene is used as a flavoring ingredient for medicinal purposes, and it has several applications in the chemical sector and home items. *Citrus* trash also contains a significant quantity of coloring pigment. They are possible sources of natural clouding agents, which are widely used in the soft drinks industry [85].

The solid residue may be used to extract essential oils, while the liquid can be utilized to make enzymes. Separation of hydrophobic substances from the skin may be used to make biodegradable polymers, packaging material, and food-grade kraft paper, reducing the requirement for petroleum-based polyesters [81].

6. *Citrus* fruit toxicity and safety profile

In animal toxicological testing, bioactive chemicals obtained from *Citrus* fruits demonstrated a high level of security. Hesperidin extracted from *C. Sinensis* dried peel revealed a modest reported adverse impact level of 1 g/kg, and a median fatal dosage of 4.83 g/kg, Sprague–Dawley rats were subjected to a 90 day subchronic and acute oral toxicity study [88]. This dose is substantially smaller than the flavonoids employed in animal research to offer the (10–200 mg/kg) same therapeutic effects, suggesting an excellent safety profile in the animals. Furthermore, additional *Citrus* flavonoids, such as tangeretin, nobiletin, and naringin, have demonstrated satisfactory effectiveness and safety [89, 90]. Limonene and other terpene-rich *Citrus* taste molecules, including oil, essential oil, peel extract, and whole fruit extract, are usually believed to be harmless [91].

7. Food-drug interaction

Grapefruit juice, for example, has an identical moniker in the context of high food-drug interaction by changing the normal working of the cytochrome oxidase mechanism. Grapefruit juice (GFJ) and HMG-CoA reductase inhibitors, sometimes referred to as statins, have the most well food-drug interaction. Grapefruit juice, when consumed in high quantities (32 oz. a day), can block the cytochrome P450 3A4 (CYP3A4) enzyme and raise serum levels of medications processed by this route, such as some statin medicines [92]. Many researchers have identified drug-GFJ interactions caused by suppression of CYP3A enzymes, a subclass of cytochrome oxidase enzyme system [93]. Furanocoumarins (in GFJ) specifically inhibit gastrointestinal CYP 3A4, increasing the oral bioavailability of medicines that are CYP 3A4 substrates, such as felodipine and cyclosporine, and eventually causing toxicity [94].

8. Future prospective

Oral supplementation delivery is acknowledged as the easiest, most efficient, and least expensive method for a broad range of medicinal significance [95]. Because of their heavy proportion in *Citrus*, bioflavonoids, such as naringenin, hesperidin, naringin, and polymethoxyflavones (PMFs) have been intensively investigated and encapsulated, and have illustrated medicinal characteristics *in vitro* and *in vivo* studies. Oral ingestion using novel encapsulating techniques, such as emulsification, liposomal interactions, hydro-gelation, nanoparticles, and membranes revolutionized the nutrition delivery mechanism [96]. Modern encapsulation techniques are developed to conserve bioflavonoids while improving target dispersion for medicinal benefits. The rapidly expanding worldwide food encapsulation business has shown higher demand and public knowledge of health additives and preventative measures for numerous diseases [92]. New unique delivery techniques must be developed to expedite the simple availability of *Citrus* bioflavonoid for medicines. Future studies,

DOI: http://dx.doi.org/10.5772/intechopen.106420

on the other hand, would need to blur the lines between the laboratory and the customer by boosting output and enhancing shelf life.

9. Conclusion

Citrus contains a variety of secondary metabolites, including phenolic acids, alkaloids, coumarins, flavonoids, carotenoids, limonoids, and volatile chemicals, which give a rational foundation for a variety of biological functions. Flavonoids, particularly methoxylated, flavones and flavanonols, and flavanones have greater bioactivities than other secondary metabolites. All of these active forms, however, operate together to give antimicrobial, antioxidative, anti-cancer, anti-inflammatory, hepatoprotection, neuroprotection, cardiac protection, and other advantages. *Citrus* are beneficial fruits to eat daily because of having multiple active forms with varied biological properties, both for their nutritive value and as chemotherapeutic or additional medication to promote health. *Citrus* fruit has various applications in the food industry. It can be used in the production of various food products. The bioactive compound, which is extracted from *Citrus* may be used in food additives, as stabilizers, thickeners, emulsifiers, preservatives, food colorants, and for many other purposes.

Acknowledgements

We are thankful to the digital library of GCUF for providing us with access to the publication.

Conflicts of interest

We have no conflict of interest.

Author details

Sakhawat Riaz[1*], Arslan Ahmad[1], Rimsha Farooq[2], Nasir Hussain[3], Tariq Riaz[4], Khadim Hussain[3] and Muhammad Mazahir[5]

1 Department of Home Economics, Government College University Faisalabad, Faisalabad, Pakistan

2 School of Food Science and Engineering, Yangzhou University, Yangzhou, China

3 Department of Food Science and technology, Government College University Faisalabad, Faisalabad, Pakistan

4 Office of the Human Resource Development Center Baltistan, Skardu, Baltistan

5 Institute of Food and Nutrition science, PMAS Arid Agriculture University Rawalpindi, Rawalpindi, Pakistan

*Address all correspondence to: sakhawat.riaz@caseit.edu.pk

References

[1] Liu Y, Heying E, Tanumihardjo SA. History, global distribution, and nutritional importance of citrus fruits. Comprehensive Reviews in Food Science and Food Safety. 2012;**11**(6):530-545. DOI: 10.1111/j.1541-4337.2012.00201.x

[2] Liu P. World markets for organic citrus and citrus juices. FAO Commodities Trade Policy and Research. 2003;**5**:1-19

[3] Ma G, Zhang L, Sugiura M, Kato M. Citrus and Health. Amsterdam: Elsevier Inc; 2020. DOI: 10.1016/B978-0-12-812163-4.00024-3

[4] Jandrić Z, Islam M, Singh DK, Cannavan A. Authentication of Indian Citrus fruit/fruit juices by untargeted and targeted metabolomics. Food Control. 2017;**72**:181-188. DOI: 10.1016/j.foodcont.2015.10.044

[5] Caristi C, Bellocco E, Panzera V, Toscano G, Vadalà R, Leuzzi U. Flavonoids detection by HPLC-DAD-MS-MS in lemon juices from sicilian cultivars. Journal of Agricultural and Food Chemistry. 2003;**51**(12):3528-3534. DOI: 10.1021/jf0262357

[6] Lv X et al. Citrus fruits as a treasure trove of active natural metabolites that potentially provide benefits for human health. Chemistry Central Journal. 2015;**9**(1):68

[7] Radhika G, Sudha V, Mohan Sathya R, Ganesan A, Mohan V. Association of fruit and vegetable intake with cardiovascular risk factors in urban south Indians. The British Journal of Nutrition. 2008;**99**(2):398-405. DOI: 10.1017/S0007114507803965

[8] Nagy S, Attaway JA. Citrus Nutrition and Quality. ACS Symposium Series. Nutrients and Nutrition of Citrus Fruits. 1980;**143**:3-24. DOI: 10.1021/bk-1980-0143

[9] Khatkar BS, Panghal A, Singh U. Applications of cereal starches in food processing. Indian food industry. 2009;**28**(2):37-44

[10] Panghal A, Kumar V, Dhull S, Gat Y, Chhikara N. Utilization of dairy industry waste-whey in formulation of papaya RTS beverage. Current Research in Nutrition and Food Science. 2017;**5**:168-174

[11] Panghal A, Janghu S, Virkar K, Gat Y, Kumar V, Chhikara N. Potential non-dairy proftic products – A healthy approach. Food Bioscience. 2018;**21**:80-89. DOI: 10.1016/j.fbio.2017.12.003

[12] Lapuente E. Relation of fruits and vegetables with major cardiometabolic risk factors, markers of oxidation, and inflammation. Nutrients. 2019;**11**:2381

[13] Zou Z, Xi W, Hu Y, Nie C, Zhou Z. Antioxidant activity of Citrus fruits. Food Chemistry. 2016;**196**:885-896. DOI: 10.1016/j.foodchem.2015.09.072

[14] Citrus: World Markets and Trade | USDA Foreign Agricultural Service. 27 July, 2022. Available from: https://www.fas.usda.gov/data/citrus-world-markets-and-trade [Accessed: 9/10/2022]

[15] Turner T, Burri BJ. Potential nutritional benefits of current Citrus consumption. Ağrı. 2013;**3**(1):170-187. DOI: 10.3390/agriculture3010170

[16] Nutritional and health benefits of Citrus fruits, 1999. ISSN: 1014-806X. Available from: https://agris.fao.org/agris-search/search.do?recordID=XF2000392903 [Accessed: 9/10/2022]

[17] YAO LH et al. Flavonoids in food and their health benefits. Plant Foods for Human Nutrition. 2004;**59**(3):113-122. DOI: 10.1007/s11130-004-0049-7

[18] Sir Elkhatim KA, Elagib RAA, Hassan AB. Content of phenolic compounds and vitamin C and antioxidant activity in wasted parts of Sudanese citrus fruits. Food Science & Nutrition. 2018;**6**(5):1214-1219

[19] Aruoma OI, Landes B, Ramful-Baboolall D, Bourdon E, Neergheen-Bhujun V, Wagner KH, et al. Functional benefits of citrus fruits in the management of diabetes. Preventive Medicine. 2012;**54**:S12-S16

[20] Peterson JJ et al. Flavanones in grapefruit, lemons, and limes: A compilation and review of the data from the analytical literature. Journal of Food Composition and Analysis. Aug. 2006;**19**:S74-S80

[21] Chen J, Montanari AM, Widmer WW. Two new polymethoxylated flavones, a class of compounds with potential anticancer activity, isolated from cold pressed dancy tangerine peel oil solids. Journal of Agricultural and Food Chemistry. 1997;**45**(2):364-368. DOI: 10.1021/jf960110i

[22] Cebadera-Miranda L et al. Sanguinello and Tarocco (Citrus sinensis [L.] Osbeck): Bioactive compounds and colour appearance of blood oranges. Food Chemistry. 2019;**270**:395-402. DOI: 10.1016/j.foodchem.2018.07.094

[23] Lado J, Gambetta G, Zacarias L. Key determinants of citrus fruit quality: Metabolites and main changes during maturation. Science Horticulture. 2018;**233**:238-248

[24] Stohs SJ, Preuss HG, Shara M. The safety of Citrus aurantium (Bitter Orange) and its primary protoalkaloid p -Synephrine. Pharmacological Research. 2011;**25**(10):1421-1428. DOI: 10.1002/ptr.3490

[25] Luan Y, Wang S, Wang R, Xu C. Accumulation of red apocarotenoid β-citraurin in peel of a spontaneous mutant of huyou (Citrus changshanensis) and the effects of storage temperature and ethylene application. Food Chemistry. 020;**309**:125705. DOI: 10.1016/j.foodchem.2019.125705

[26] Peng Z et al. Comparative profiling and natural variation of polymethoxylated flavones in various Citrus germplasms. Food Chemistry. 2021;**354**:129499. DOI: 10.1016/j.foodchem.2021.129499

[27] Breithaupt DE, Bamedi A. Carotenoid esters in vegetables and fruits: A screening with emphasis on β-cryptoxanthin esters. Journal of Agricultural and Food Chemistry. 2001;**49**(4):2064-2070. DOI: 10.1021/jf001276t

[28] Saini RK, Ranjit A, Sharma K, Prasad P, Shang X, Gowda KGM, et al. Bioactive compounds of citrus fruits: A review of composition and health benefits of carotenoids, flavonoids, limonoids, and terpenes. Antioxidants. 2022;**11**(2):239

[29] Dhuique-Mayer C, Gence L, Portet K, Tousch D, Poucheret P. Preventive action of retinoids in metabolic syndrome/type 2 diabetic rats fed with Citrus functional food enriched in β-cryptoxanthin. Food & Function. 2020;**11**(10):9263-9271. DOI: 10.1039/D0FO02430A

[30] Hemanth Kumar B, Dinesh Kumar B, Diwan PV. Hesperidin, a Citrus flavonoid, protects against L-methionine-induced hyperhomocysteinemia by abrogation of oxidative stress, endothelial dysfunction and neurotoxicity in Wistar rats.

Pharmaceutical Biology. 2017;**55**(1):146-155. DOI: 10.1080/13880209.2016.1231695

[31] Ciftci O, Ozcan C, Kamisli O, Cetin A, Basak N, Aytac B. Hesperidin, a Citrus flavonoid, has the ameliorative effects against experimental autoimmune Encephalomyelitis (EAE) in a C57BL/J6 Mouse Model. Neurochemical Research. 2015;**40**(6):1111-1120. DOI: 10.1007/s11064-015-1571-8

[32] Roohbakhsh A, Parhiz H, Soltani F, Rezaee R, Iranshahi M. Molecular mechanisms behind the biological effects of hesperidin and hesperetin for the prevention of cancer and cardiovascular diseases. Life Sciences. 2015;**124**:64-74. DOI: 10.1016/j.lfs.2014.12.030

[33] Singh J, Sood S, Muthuraman A. In-vitro evaluation of bioactive compounds, anti-oxidant, lipid peroxidation and lipoxygenase inhibitory potential of Citrus karna L. peel extract. Journal of Food Science and Technology. 2014;**51**(1):67-74. DOI: 10.1007/s13197-011-0479-9

[34] Padilla-Camberos E, Lazcano-Díaz E, Flores-Fernandez JM, Owolabi MS, Allen K, Villanueva-Rodríguez S. Evaluation of the inhibition of carbohydrate hydrolyzing enzymes, the antioxidant activity, and the polyphenolic content of Citrus limetta peel extract. Scientific World Journal. 2014;**2014**:1-4. DOI: 10.1155/2014/121760

[35] Heiss E, Herhaus C, Klimo K, Bartsch H, Gerhäuser C. Nuclear factor κB is a molecular target for sulforaphane-mediated anti-inflammatory mechanisms. The Journal of Biological Chemistry. 2001;**276**(34):32008-32015. DOI: 10.1074/jbc.M104794200

[36] Grosso G, Bei R, Mistretta A, Marventano S, Calabrese G, Masuelli L, et al. Effects of vitamin C on health: A review of evidence. Frontiers in Bioscience. 2013;**18**:1017-1029

[37] Bellavite P, Donzelli A. Hesperidin and SARS-CoV-2: New light on the healthy function of Citrus fruits. Antioxidants. 2020;**9**(8):1-18. DOI: 10.3390/antiox9080742

[38] Wallace TC et al. Fruits, vegetables, and health: A comprehensive narrative, umbrella review of the science and recommendations for enhanced public policy to improve intake. Critical Reviews in Food Science and Nutrition. 2020;**60**(13):2174-2211. DOI: 10.1080/10408398.2019.1632258

[39] Barreca D, Mandalari G, Calderaro A, Smeriglio A, Trombetta D, Felice MR, et al. Citrus flavones: An update on sources, biological functions, and health promoting properties. Plants. 2020;**9**(3):288

[40] Zhu C, Zhou X, Long C, Du Y, Li J, Yue J, et al. Variations of flavonoid composition and antioxidant properties among different cultivars, fruit tissues and developmental stages of citrus fruits. Chemistry & Biodiversity. 2020;**17**(6):e1900690

[41] Wu C et al. Analysis of therapeutic targets for SARS-CoV-2 and discovery of potential drugs by computational methods. Acta Pharmaceutica Sinica B. 2020;**10**(5):766-788. DOI: 10.1016/j.apsb.2020.02.008

[42] Chen YW, Yiu C-PB, Wong K-Y. Prediction of the SARS-CoV-2 (2019-nCoV) 3C-like protease (3CLpro) structure: Virtual screening reveals velpatasvir, ledipasvir, and other drug repurposing candidates. F1000Research. 2020;**9**:129

[43] Adem S, Eyupoglu V, Sarfraz I, Rasul A, Ali M. Identification of Potent

COVID-19 Main Protease (Mpro) Inhibitors from Natural Polyphenols: An in Silico Strategy Unveils a Hope against CORONA. Preprints. 2020:2020030333. DOI: 10.20944/preprints202003.0333.v1

[44] Das S, Sarmah S, Lyndem S, Singha Roy A. An investigation into the identification of potential inhibitors of SARS-CoV-2 main protease using molecular docking study. Journal of Biomolecular Structure & Dynamics. 2020;**2020**:1-11. DOI: 10.1080/07391102.2020.1763201

[45] Joshi RS et al. Discovery of potential multi-target-directed ligands by targeting host-specific SARS-CoV-2 structurally conserved main protease. Journal of Biomolecular Structure & Dynamics. 2020;**2020**:1-16. DOI: 10.1080/07391102.2020.1760137

[46] Park H-K, Kang SW, Park M-S. Hesperidin ameliorates hepatic ischemia-reperfusion injury in sprague-dawley rats. Transplantation Proceedings. 2019;**51**(8):2828-2832. DOI: 10.1016/j.transproceed.2019.02.059

[47] Jo SH et al. Hesperetin inhibits neuroinflammation on microglia by suppressing inflammatory cytokines and MAPK pathways. Archives of Pharmacal Research. 2019;**42**(8):695-703. DOI: 10.1007/s12272-019-01174-5

[48] Miler M et al. Citrus flavanones naringenin and hesperetin improve antioxidant status and membrane lipid compositions in the liver of old-aged Wistar rats. Experimental Gerontology. 2016;**84**:49-60. DOI: 10.1016/j.exger.2016.08.014

[49] Messina G, Polito R, Monda V, Cipolloni L, Di Nunno N, Di Mizio G, et al. Functional role of dietary intervention to improve the outcome of COVID-19: A hypothesis of work. International Journal of Molecular Sciences. 2020;**21**(9):3104

[50] Kalantar-Zadeh K, Moore LW. Impact of nutrition and diet on COVID-19 infection and implications for kidney health and kidney disease management. Journal of Renal Nutrition. 2020;**30**(3):179-181

[51] Iranshahi M, Rezaee R, Parhiz H, Roohbakhsh A, Soltani F. Protective effects of flavonoids against microbes and toxins: The cases of hesperidin and hesperetin. Life Sciences. 2015;**137**:125-132. DOI: 10.1016/j.lfs.2015.07.014

[52] Grosso G et al. Red orange: Experimental models and epidemiological evidence of its benefits on human health. Oxidative Medicine and Cellular Longevity. 2013;**2013**:1-11. DOI: 10.1155/2013/157240

[53] Siahpoosh A, Javedani F. Antioxidative capacity of Iranian Citrus deliciosa peels. Free Radicals and Antioxidants. 2012;**2**(2):62-67. DOI: 10.5530/ax.2012.2.2.11

[54] Hirata T et al. Identification and physiological evaluation of the components from Citrus fruits as potential drugs for anti-corpulence and anticancer. Bioorganic & Medicinal Chemistry. 2009;**17**(1):25-28. DOI: 10.1016/j.bmc.2008.11.039

[55] Nicole Cotelle BSP. Role of flavonoids in oxidative stress. Current Topics in Medicinal Chemistry. 2001;**1**(6):569-590. DOI: 10.2174/1568026013394750

[56] Marí M, Colell A, Morales A, von Montfort C, Garcia-Ruiz C, Fernández-Checa JC. Redox control of liver function in health and disease. Antioxidants & Redox Signaling. 2010;**12**(11):1295-1331. DOI: 10.1089/ars.2009.2634

[57] Shirani K, Yousefsani BS, Shirani M, Karimi G. Protective effects of naringin

against drugs and chemical toxins induced hepatotoxicity: A review. Pharmacological Research. 2020;**34**(8):1734-1744. DOI: 10.1002/ptr.6641

[58] Yoshizaki N, Fujii T, Masaki H, Okubo T, Shimada K, Hashizume R. Orange peel extract, containing high levels of polymethoxyflavonoid, suppressed UVB-induced COX-2 expression and PGE 2 production in HaCaT cells through PPAR- γ activation. Experimental Dermatology. 2014;**23**:18-22. DOI: 10.1111/exd.12394

[59] Kwon KH, Murakami A, Tanaka T, Ohigashi H. Dietary rutin, but not its aglycone quercetin, ameliorates dextran sulfate sodium-induced experimental colitis in mice: Attenuation of pro-inflammatory gene expression. Biochemical Pharmacology. 2005;**69**(3):395-406

[60] Dodda D, Chhajed R, Mishra J, Padhy M. Targeting oxidative stress attenuates trinitrobenzene sulphonic acid induced inflammatory bowel disease like symptoms in rats: Role of quercetin. Indian Journal of Pharmacology. 2014;**46**(3):286

[61] Cibiček N, Roubalová L, Vrba J, Zatloukalová M, Ehrmann J, Zapletalová J, et al. Protective effect of isoquercitrin against acute dextran sulfate sodium-induced rat colitis depends on the severity of tissue damage. Pharmacological Reports. 2016;**68**(6):1197-1204

[62] Kumar VS, Rajmane AR, Adil M, Kandhare AD, Ghosh P, Bodhankar SL. Naringin ameliorates acetic acid induced colitis through modulation of endogenous oxido-nitrosative balance and DNA damage in rats. Journal of Biomedical Research. 2014;**28**(2):132-145

[63] Hoensch HP, Weigmann B. Regulation of the intestinal immune system by flavonoids and its utility in chronic inflammatory bowel disease. World Journal of Gastroenterology. 2018;**24**(8):877-881

[64] Benavente-García O, Castillo J. Update on uses and properties of Citrus flavonoids: New findings in anticancer, cardiovascular, and anti-inflammatory activity. Journal of Agricultural and Food Chemistry. 2008;**56**(15):6185-6205

[65] Nunes C, Almeida L, Barbosa RM, Laranjinha J. Luteolin suppresses the JAK/STAT pathway in a cellular model of intestinal inflammation. Food & Function. 2017;**8**(1):387-396

[66] Xiong Y, Chen D, Yu C, Lv B, Peng J, Wang J, et al. Citrus nobiletin ameliorates experimental colitis by reducing inflammation and restoring impaired intestinal barrier function. Molecular Nutrition & Food Research. 2015;**59**(5):829-842

[67] Tanaka T, Tanaka T, Tanaka M, Kuno T. Cancer chemoprevention by Citrus pulp and juices containing high amounts of β-cryptoxanthin and hesperidin. Journal of Biomedicine & Biotechnology. 2012;**2012**:1-10. DOI: 10.1155/2012/516981

[68] Xiao H, Yang CS, Li S, Jin H, Ho C-T, Patel T. Monodemethylated polymethoxyflavones from sweet orange (Citrus sinensis) peel Inhibit growth of human lung cancer cells by apoptosis. Molecular Nutrition & Food Research. 2009;**53**(3):398-406. DOI: 10.1002/mnfr.200800057

[69] JIN H et al. Flavonoids from Citrus unshiu Marc. inhibit cancer cell adhesion to endothelial cells by selective inhibition of VCAM-1. Oncology Reports. 2013;**30**(5):2336-2342. DOI: 10.3892/or.2013.2711

[70] Lai C-S et al. Effective suppression of azoxymethane-induced aberrant crypt foci formation in mice with Citrus peel flavonoids. Molecular Nutrition & Food Research. 2013;**57**(3):551-555. DOI: 10.1002/mnfr.201200606

[71] Chang L et al. Ougan (Citrus reticulata cv. Suavissima) flavedo extract suppresses cancer motility by interfering with epithelial-to-mesenchymal transition in SKOV3 cells. China Medical. 2015;**10**(1):14

[72] Ravishankar D, Rajora AK, Greco F, Osborn HMI. Flavonoids as prospective compounds for anti-cancer therapy. The International Journal of Biochemistry & Cell Biology. 2013;**45**(12):2821-2831

[73] Visalli G, Ferlazzo N, Cirmi S, Campiglia P, Gangemi S, Di Pietro A, et al. Bergamot juice extract inhibits proliferation by inducing apoptosis in human colon cancer cells. Anti-Cancer Agents in Medicinal Chemistry. 2014;**14**:1402-1413

[74] Mink PJ et al. Flavonoid intake and cardiovascular disease mortality: A prospective study in postmenopausal women. The American Journal of Clinical Nutrition. 2007;**85**(3):895-909. DOI: 10.1093/ajcn/85.3.895

[75] Lin Y et al. Molecular structures of Citrus flavonoids determine their effects on lipid metabolism in HepG2 cells by primarily suppressing ApoB secretion. Journal of Agricultural and Food Chemistry. 2011;**59**(9):4496-4503. DOI: 10.1021/jf1044475

[76] Demonty I et al. The Citrus flavonoids hesperidin and naringin do not affect serum cholesterol in moderately hypercholesterolemic men and women. The Journal of Nutrition. 2010;**140**(9):1615-1620. DOI: 10.3945/jn.110.124735

[77] Shen W, Xu Y, Lu Y-H. Inhibitory effects of Citrus flavonoids on starch digestion and antihyperglycemic effects in HepG2 Cells. Journal of Agricultural and Food Chemistry. 2012;**60**(38):9609-9619. DOI: 10.1021/jf3032556

[78] Akiyama S, Katsumata S, Suzuki K, Ishimi Y, Wu J, Uehara M. Dietary hesperidin exerts hypoglycemic and hypolipidemic effects in streptozotocin-induced marginal type 1 diabetic rats. Journal of Clinical Biochemistry and Nutrition. 2009;**46**(1):87-92. DOI: 10.3164/jcbn.09-82

[79] Annadurai T, Muralidharan AR, Joseph T, Hsu MJ, Thomas PA, Geraldine P. Antihyperglycemic and antioxidant effects of a flavanone, naringenin, in streptozotocin–nicotinamide-induced experimental diabetic rats. Journal of Physiology and Biochemistry. 2012;**68**(3):307-318. DOI: 10.1007/s13105-011-0142-y

[80] Roghani M, Fallahi F, Moghadami S. Citrus flavonoid naringenin improves aortic reactivity in streptozotocin-diabetic rats. Indian Journal of Pharmacology. 2012;**44**(3):382

[81] Chavan P, Singh AK, Kaur G. Recent progress in the utilization of industrial waste and by-products of Citrus fruits: A review. Journal of Food Process Engineering. 2018;**41**(8):1-10. DOI: 10.1111/jfpe.12895

[82] Taghizadeh-Alisaraei A, Hosseini SH, Ghobadian B, Motevali A. Biofuel production from Citrus wastes: A feasibility study in Iran. Renewable and Sustainable Energy Reviews. 2017;**69**:1100-1112. DOI: 10.1016/j.rser.2016.09.102

[83] Casquete R et al. High pressure extraction of phenolic compounds from Citrus peels†. High Pressure Research. 2014;**34**(4):447-451

[84] Marangoni AG. Editorial overview: Food chemistry and biochemistry. Current Opinion in Food Science. 2016;7:iv-v

[85] Mahato N, Sharma K, Sinha M, Cho MH. Citrus waste derived nutra-/ pharmaceuticals for health benefits: Current trends and future perspectives. Journal of Functional Foods. 2018;**40**:307-316. DOI: 10.1016/j.jff.2017.11.015

[86] Piriyaprasarth S, Sriamornsak P. Flocculating and suspending properties of commercial Citrus pectin and pectin extracted from pomelo (Citrus maxima) peel. Carbohydrate Polymers. Jan. 2011;**83**(2):561-568. DOI: 10.1016/j.carbpol.2010.08.018

[87] Maria Galati E et al. Anti-inflammatory effect of lemon mucilage: In vivo and in vitro studies. Immunopharmacology and Immunotoxicology. 2005;**27**(4):661-670. DOI: 10.1080/08923970500418919

[88] Li Y, Kandhare AD, Mukherjee AA, Bodhankar SL. Acute and sub-chronic oral toxicity studies of hesperidin isolated from orange peel extract in Sprague Dawley rats. Regulatory Toxicology and Pharmacology. 2019;**105**:77-85. DOI: 10.1016/j.yrtph.2019.04.001

[89] Li P et al. Toxicological evaluation of Naringin: Acute, subchronic, and chronic toxicity in Beagle Dogs. Regulatory Toxicology and Pharmacology. 2020;**111**:104580

[90] Nakajima A, Nemoto K, Ohizumi Y. An evaluation of the genotoxicity and subchronic toxicity of the peel extract of Ponkan Cultivar 'Ohta Ponkan' (Citrus Reticulata Blanco) that is rich in nobiletin and tangeretin with anti-dementia activity. Regulatory Toxicology and Pharmacology. 2020;**114**:104670

[91] Cohen SM et al. FEMA GRAS assessment of natural flavor complexes: Citrus-derived flavoring ingredients. Food and Chemical Toxicology. 2019;**124**:192-218. DOI: 10.1016/j.fct.2018.11.052

[92] Kafle A, Mohapatra SS, Sarma J, Reddy I. Food-drug interaction: A review. Pharma Innovation Journal. 2018;7(1):114-118

[93] Pawełczyk T, Kłoszewska I. Grapefruit juice interactions with psychotropic drugs: Advantages and potential risk. Przegląd Lekarski. 2008;**65**(2):92-95

[94] de Castro WV, Mertens-Talcott S, Derendorf H, Butterweck V. Grapefruit juice drug interactions: Grapefruit juice and its components inhibit Pglycoprotein (ABCB1) mediated transport of talinolol in Caco-2 cells. Journal of Pharmaceutical Sciences. 2007;**96**(10):2808-2817

[95] Ghosh T, Ghosh A, Prasad D. A review on new generation orodispersible tablets and its future prospective. International Journal of Pharmacy and Pharmaceutical Sciences. 2011;**3**(1):1-7

[96] Tibbitt MW, Dahlman JE, Langer R. Emerging frontiers in drug delivery. Journal of the American Chemical Society. 2016;**138**(3):704-717. DOI: 10.1021/jacs.5b09974

Chapter 10

Inhibition of Multidrug Resistance by Polyphenolic Phytochemicals of Citrus Fruits

Anne Adebukola Adeyanju

Abstract

The incidence of multidrug resistance (MDR) during treatment is rising, persisting and spreading globally due to the emergence of resistance to multiple antibiotics. This is a serious challenge that is causing clinical failures in the treatment of infections and diseases, persistent illnesses, higher costs for health maintenance, higher risk of mortality and threat to public health. It is widely acknowledged that polyphenols present in citrus fruits exhibit beneficial effects in the inhibition of MDR, and as such, they could serve as appealing potential therapeutics in this regard. Some of the mechanisms citrus polyphenols employed in preventing the occurrence of MDR in cancer and bacterial cells have been identified. They include the initiation of apoptosis through modulation of the PI3K/Akt pathway, inhibition of P-glycoprotein function, enhancement of the intracellular accumulation of drugs, arrest of the cell cycle at G0/G1 phase, downregulation of the expression of pro-inflammatory cytokines and others. Taking into account the significance of MDR, this review highlights the mechanisms of MDR and the potential inherent in citrus polyphenols in fruits in reversing it.

Keywords: multidrug resistance, citrus, polyphenol, signaling pathway, flavonoids

1. Introduction

Citrus belongs to one of the most popular and important fruit crops that is widely grown in the world and supplies valuable and essential nutrients to the human diet. Citrus fruits depict a rich origin of naturally occurring biologically active compounds which are polyphenolic phytochemicals. Their abundant level of flavonoids, terpenes, phytonutrients and other phenolic compounds, carotenoids and vitamins C have been described [1]. Polyphenols are one of the major functional bioactive compounds available in citrus such as grapefruits (*Citrus paradisi*), oranges (*Citrus sinensis*), lime (*Citrus aurantifolia*), tangerine (*Citrus reticulata*) and others. Consumption of polyphenols of more than 820 mg daily has been identified in several dietary sources including fruits and vegetables. Gradually, the consideration of polyphenolic compounds because of their therapeutic benefit is coming to play, as the world returns to nature to discover sources of prospective drug candidates [2, 3].

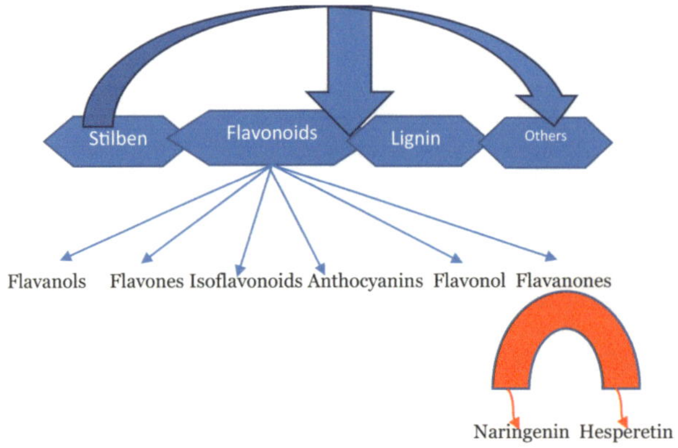

Figure 1.
Commonly found polyphenols in citrus fruits.

Polyphenol represents a chemical compound with a structure that consists of one or more phenolic rings [4]. Further, classifications are done based on the number of phenolic rings and other structures associated with these rings (for example, oxygenated heterocycle) [5]. Polyphenols are most recognized for possession of unique structural features which include the flavan ring and many units of phenol. They are beneficial organic compounds placed into different subclasses (**Figure 1**). Currently, nearly 15,000 varieties have been recognized, wherein flavonoids represent one of the polyphenols [2]. Flavonoids' structure contains 15 carbon atoms that consist of two benzene rings, connected to a flavone ring. Oxidation of the heterocyclic ring can give rise to other subclasses such as flavanones, flavonols, flavones and anthocyanins. Citrus fruits such as lemons (*Citrus limon*), grapefruit, oranges and limes are sometimes referred to as citroflavonoids due to their richness in flavanones [6].

Multidrug resistance (MDR) of cancer and bacteria cells constitutes one of the main challenges and an obstacle that is usually encountered during chemotherapy, and this often results in chemotherapy failure [7]. Thus, the search for non-toxic modulators of MDR is of great importance and beneficial effect on enhancing the apoptosis of resistant cells. Even the rapid increase of microorganisms' resistance to antimicrobial drugs is currently a task to tackle, globally [8]. The emergence of bioactive compounds found in nature among plants or their products has been attracting the curiosity of researchers with a focus on the biological significance of citrus fruits as a possible and prospective source of pharmaceuticals [3].

Recently, studies have demonstrated the involvement of the mobilization of Phosphoinositide 3-kinase/protein kinase (PI3K/AKT), extracellular-signal-regulated kinase (ERK) and nuclear factor erythroid 2–related factor 2 (Nrf2) pathways in resisting the effects of chemotherapeutic drugs [9, 10].

Since anti-tumor drugs can inhibit these signaling pathways for increased sensitivity of tumor cells to chemotherapy drugs [11], identification of potent and potential inhibiting agents that can inhibit these pathways for reversal of MDR is necessary. Here, we discuss the consideration of natural products such as polyphenolic phytochemicals found in citrus fruits as prospective MDR modifiers since they are safe and harmless [12, 13].

2. Polyphenols in citrus fruits (classifications, structures and beneficial effects)

Polyphenols are a large family of thousands of plant compounds that have been identified for their unique and characteristic structures which include a flavan ring system and many units of phenol [4, 5]. Polyphenols are classified majorly into flavonoids and non-flavonoids. Flavonoids are a class of polyphenols that have been well researched. They share a common flavan nucleus, which consists of a pair of aromatic rings A and B, aromatic and connected by a pyran ring (C). The location difference of the B-ring to C-ring linkage enables the classification among flavonoids, isoflavonoids and neoflavonoids (4-phenylbenzopyrans). Flavonoids, which are the most abundant may be also categorized into flavonols, dihydroflavonols, anthocyanidins, flavanols and flavonoids with no substitution at C3 (**Figure 2**). The detection of a double bond at the position of C2–C3 defines the difference between flavones and flavanones [14].

Flavanones could be considered the main substrate for the biosynthesis of flavonoids as they serve as the precursors of other flavonoid classes. Naringenin and hesperetin (with their aglycones and their glycosides) are flavanones that have sparked great curiosity and interest due to their abundant occurrence in foods. Hesperidin is an important flavanone contained in every citrus fruit especially, with good concentrations in sweet oranges [15] and orange juice [16]. Naringenin-7-*O*-rutinoside (narirutin) is also found in orange juices with traces of

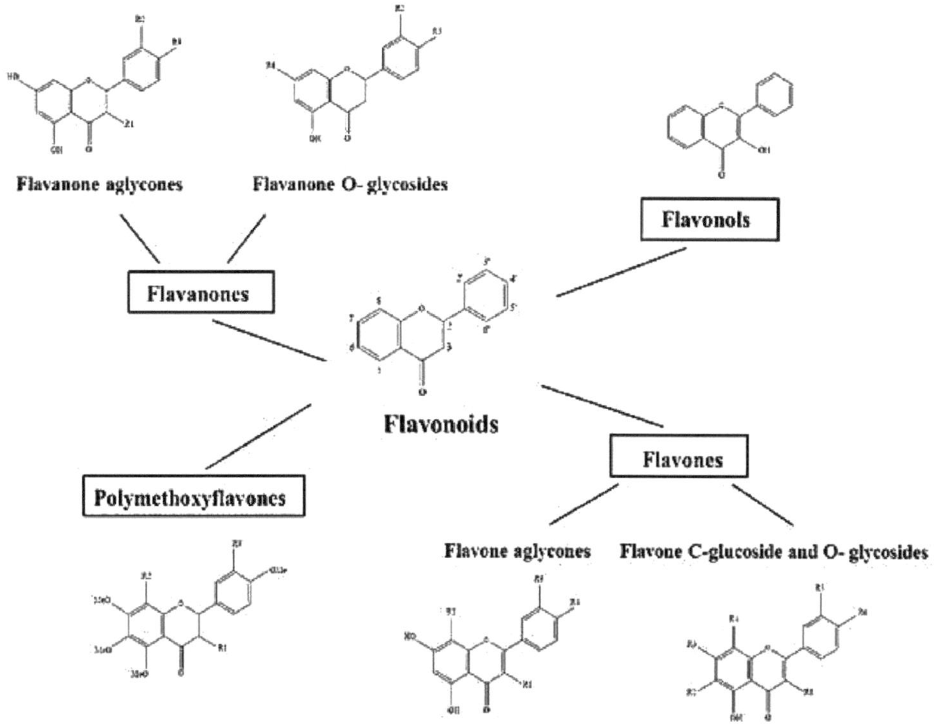

Figure 2.
Structures of citrus flavonoid subclasses.

hesperetin-7-*O*-rutinoside-3′-*O*-glucoside, 4′-*O*-methyl-naringenin-7-*O*-rutinoside (didymin) and eriodictyol-7-*O*-rutinoside (eriocitrin) [17].

Naringenin, structurally recognized as 5,7,40 -trihydroxyflavanone, is present and largely concentrated in citrus fruits [18]. Naringenin can occur as aglycone and/or glycosides where naringin and narirutin are most abundant. Naringin also known as naringenin-7- neohesperidoside has a characteristic bitter flavor because of the glucose component. It is the main flavonoid found in grape and sour oranges [19], with varying naringin contents according to their varieties. However, other citrus species such as tangelo (*C. reticulata x C. paradisi*), lemon, sweet oranges and lime have a low content of naringin. Narirutin, also known as naringenin-7-rutinoside, is a major naringenin glycoside. It is very abundant in grapefruit although not as much as naringin. At some detectable levels, narirutin is also found in citrus fruits such as sweet oranges, tangor, tangelo and tangerine [20]. During processing, the naringenin content of these fruits may be affected due to the liberation and degradation of associated flavanones because of their thermo-responsive nature [21].

Hesperetin together with the glycosides is found in citrus fruits [22]. Its glycosides form is more dominant than the aglycone. Hesperidin and neohesperidin are the most widely distributed glycosides and are conjugates of rutinose and neohesperidose, respectively. Hesperidin exists in large amounts in sweet oranges, grapefruits, limes, lemons, tangerine and tangor (*C. reticulata* × *C. sinensis*) [22]. Neohesperidin is highly present in a significant amount in tangelo and sour orange.

Polyphenols in citrus have a lot of beneficial effects. Their free radical scavenging capacity has been reported *in vitro* [23] while their potential to activate the endogenous antioxidant defense system has been described. The activation was found to coincide with depletion in the level of reactive oxygen molecules such as nitric oxide ($NO^{\cdot}$), hydrogen peroxide (H_2O_2) and other biological markers of oxidative stress. In addition, the involvement of overexpression of the transcription factor Nrf2 [24, 25] has been linked to the mediated stimulation of the antioxidant response element. Studies by [26] showed the upregulation of haem-oxygenase (HO-1) and a decrease in the activity of xanthine oxidase (XO), a superoxide radical generating enzyme by hesperidin and hesperetin.

Citrus polyphenols also possess immunomodulatory activities [26]. The molecular targets involve the lowering of the levels of pro-inflammatory cytokines *interleukin*-1β (IL-1β), *interleukin*-2 (IL-2), *interleukin*-6 (IL-6), interferon-γ (IFN-γ) and tumor necrosis factor-α (TNF-α) which is probably facilitated through the weakening of overactive immune cells as is demonstrated by the fall in glial fibrillary acidic protein (GFAP) level and reduction of nuclear factor kappa-light-chain-enhancer of activated B cells (NF-κB) which controls the transcription of chemokines and inflammatory mediators [27, 28]. Their ability to ameliorate mitochondrial dysfunction has been described [29] as this can increase the stability and function of mitochondrial respiratory chain complexes (I–IV).

3. Multidrug resistance: a looming global crisis

MDR can be defined as insensitivity to administered drugs despite earlier responsiveness to them [30, 31]. Microorganisms such as viruses, fungi, bacteria and parasites can be resistant to antimicrobial drugs, leading to unsuccessful treatment and resulting in a continuous spread of infections [32]. MDR promotes difficulty in disease management by elevating the likelihood of the spread of resistant pathogens.

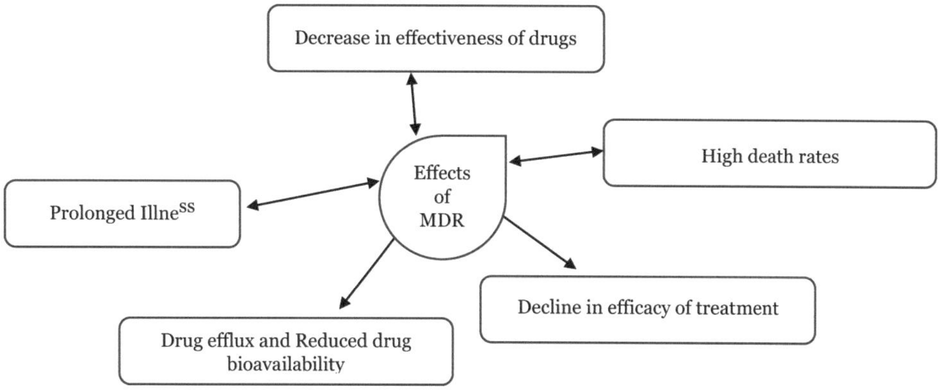

Figure 3.
Effects of multidrug resistance.

This declines the efficacy of treatment, thereby extending the time of infection in patients (**Figure 3**). Further, the adoption of expensive therapies is also triggered by the rise in cost implications of treatment attributed to MDR as the pathogens would have turned out to be insusceptible to the accessible drugs. In addition, there is suppressed immune function and decreased drug bioavailability due to the inflated rate of drug metabolism [32].

The emergence of infectious microorganisms having a considerable number of resistant species and the ability to counter barriers imposed by drugs during treatment has unfolded over the years, and this is quite alarming. It's well documented that virtually all the known and capable infecting agents (fungi, bacteria, viruses and parasites) have employed a high level of MDR with an increased death rate [32]. Many deadly diseases such as HIV, influenza, tuberculosis, pneumonia, malaria and yeast infections, caused by these infecting agents, have been identified as major causes of death in recent times [33]. This is a global concern as MDR continues to pose a serious risk to the health of the public. For instance, there has been a decreased chance of managing tuberculosis owing to its resistance to antibiotic treatment. Likewise, pneumonia is an infection that has become untreatable because of the development of resistance to cephalosporin in response to elongated-spectrum β-lactamases-mediated mechanism [33], hence making all accessible treatment with β- lactam antibiotics useless. Drug resistance during HIV treatment has also resulted in antiretroviral therapy failure. Protozoan parasites responsible for malaria are showing resistance to most of their effective anti-malaria drugs such as chloroquine and artemisinin [34] which have demanded their replacement with novel drugs, thereby increasing healthcare expenses and contributing to the global economic burden. MDR explains why some microbes are sometimes irresponsive to treatment with standard drugs, thus, stretching treatment duration and further raising the healthcare costs, especially for those who are not capable of such expenses [35].

A series of mechanisms to survive, be resistant and subdue the effects of drugs and exposure to drugs has evolved in microorganisms. Notably, the cell walls of these microorganisms play a vital role in their viability. They strategized by undergoing mutation in their chromosomes [36], and this could result in the modification in the cell membrane structure translating to reduced penetrability and drug absorption into the cell [30]. In addition, active target binding sites for drugs may become unavailable. Though drugs for treating viral infection normally select the DNA polymerase of the virus due to their ability to reverse the transcriptase activity and prevent viral replication, drug-resistant variants of microorganisms usually undergo mutations where the reverse transcriptase is located on the polymerase gene [36]. This influences the cooperation between the drug and the enzyme.

Another mechanism of MDR in microorganisms which can affect a drug's access to the target sites is by overexpressing the target enzymes which can cause modification in some metabolic pathways, generating other target molecules and tampering with protein production [37]. This process can negatively affect the entry of drugs to the designated sites [32].

Microorganisms also exhibit MDR by inactivating or enzymatically degrading antimicrobial drugs by hydrolysing their ester or amide bonds. Other apparent ways by which drugs can be transformed by microorganisms include phosphorylation, adenylation, hydroxylation, acetylation and glycosylation [38]. They can alter the activity of antimicrobial drugs and hinder their connection with the target sites [38]. They can also induce the expression of P-glycoprotein and multidrug-resistant proteins, an action that affects the fluidity and porosity of their membranes, causing ATP-dependent outflow of the antimicrobial drugs and reducing their intracellular level [39].

One of the key challenges in subduing the existence of cancer is MDR. Some mechanisms in cancer have been linked to MDR [40]. They include target changes in enzymes such as DNA topoisomerases [41], mitotic arrest [42], interference in DNA repair [43] and apoptosis impairment or genes involved in apoptosis and necrosis [44].

A common biochemical mechanism of MDR is drug efflux via ATP-binding cassette transporters (ABC transporters). The exaggerated expression of these transporters has been connected with chemoresistance in cancer cells [45]. There are 48 members in the family in humans [46], and the most extensively studied include P-Glycoprotein (P-gp), multidrug resistance protein 1 (MRP1) and breast cancer resistance protein (BCRP). They are largely present in healthy cells of mammalian tissues for the translocation of small compounds. Their presence in the epithelial cells of the intestine and kidney proximal tubules and endothelial cells of blood capillaries has been documented [46, 47]. Expression of ABC transporters by cells can occur if anticancer drugs, which are also substrates of these transporters, are regarded as foreign. In addition, cancer cells that have no resistance to anticancer agents initially can also acquire the potential by overproducing the transporters to counter the effect of toxic substances. Eventually, this triggers an elevated efflux, lowering intracellular drug concentration triggers an elevated efflux, lowering intracellular drug concentration [48] that is deficient in killing a cancer cell.

The phosphatidylinositol 3-kinase/protein kinase B (PI3K/AKT) pathway is another signaling pathway that promotes drug efflux through the expression of ABC transporter. The pathway encourages the proliferation of tumor by activating NF-κB which can rescue the cancer cells from cell death [49], and suppression of caspase-3 activity, thereby inhibiting apoptosis. This has been considered an underlining mechanism for MDR in tumor cells.

4. Multidrug resistance inhibition mechanisms of Citrus polyphenols

4.1 In Cancer

The existence of MDR has been a notable challenge during chemotherapeutic treatment. Extensive studies have been carried out on potential MDR reversers, and several inhibitors of P-gp have been identified among the accessible drugs. However, their toxicity profiles and drug-interaction effects have propelled researchers to explore novel compounds with mild toxicity and lesser side effects [50].

Currently, the discovery of non-toxic, more efficacious and low-cost compounds originating from natural sources to tackle MDR is receiving recognition. The consumption of citrus fruits has been observed to be useful for the reduction of MDR usually experienced during cancer chemotherapy (**Table 1**). The concentration range 6.71–11.43 μM of nobiletin (also known as 5,6,7,8,3′,4′-hexamethoxyflavone) found in orange juice [55] was discovered to be sufficient for inhibition of P-glycoprotein function as documented by [56]. This polymethoxyflavonoid has been identified as a substrate of P-glycoprotein and can competitively connect at its drug-binding site. This action leaves little or no space for related compounds to attach to the transporter, thus resulting in a declined activity of ABCB1 transporter. Part of its inhibitory mechanism against MDR is on P-glycoprotein function which is dependent on ATP hydrolysis. It is known that hydrolysis of ATP is tightly coupled to the transportation of compounds by ABC transporters, with stimulation of ATPase activity [57]. Since the energy consumed by ABCB1 transporter is derived from ATP hydrolysis, nobiletin lowers the ATPase activity [58], thereby modulating its drug transport activity. This is one of the mechanisms of reversing the overexpression of ABCB1 during MDR.

The enhancement of the intracellular accumulation of drugs during chemotherapy is another mechanism of MDR inhibition by citrus polyphenols. Nobiletin at 50 μ M was discovered to have the capacity to increase daunorubicin accumulation in KB-C2 cells [56] and the absorption of vinblastine in Caco-2 cells [59] together with ABCB1 transfected LLC-GA5-COL300 cells at 20 μ M [60] demonstrating the possible P-gp inhibition action of nobiletin. Further, nobiletin at concentrations of 0.5–9 μM remarkably raised the sensitivity of ABCB1 involved in the overexpression of A2780/T and A549/T cell lines to chemotherapeutic agents [58]. Also, tangeretin is

Citrus polyphenol	Type of infection/ Cell line	Concentration	Mechanism of action	Reference
Hesperitin	Breast cancer/ MCF-7 DOX-resistant cells	0.5–3.5 μm/l	Decreased P-gp expression	[47]
	Helicobacter Pylori	4 μg/ml	Inhibition of HsrA, invasion of lipid bilayer of cell membrane	[51]
Naringenin	Helicobacter Pylori	128 μg/ml	Inhibition of urease activity	[52]
Tangeretin	Colon Cancer/ LoVo/Dx cells	75 and 100 μM	Induction of apoptosis, decrease of mitochondrial potential and caspase activation	[53]
Apigenin	W480 cells, HT-29 cells, and Caco-2 cells	80 μM	G2/M arrest	[54]
Nobiletin	KB-C2 cells	50 μM	Inhibition of P-gp, enhancement of drug accumulation	[46]

Table 1.
Mechanisms of action of some citrus polyphenols in the inhibition of MDR in cancer and bacteria cells.

shown to be an effective MDR modulator by increasing the awareness of cancer cells to doxorubicin. Another citrus polyphenol, hesperidin, likewise utilizes the sensitivity tool in cancer cells with resistance, thereby decreasing the expression of P-gp [61]. Its inhibitory effect in overcoming MDR in cancer cells is significant and higher than nobiletin [62].

The initiation of apoptosis is another mechanism citrus polyphenols employ in preventing the occurrence of MDR during chemotherapy without developing toxicity. It involves modulation of the PI3K/Akt pathway which is beneficial in preventing drug resistance in many cancer models. Hesperitin has no report of significant toxicity to normal cells [63], and its anti-tumor activity occurs through interaction with various carcinogenic signaling pathways and promoting mitochondrial apoptosis pathways. [64] described the promotion of apoptosis in gastric cancer by hesperetin. The mechanism involves the inhibition of the PI3K/AKT signaling pathway through the upregulation of phosphatase and tensin homolog (PTEN) expression, thus hampering cell proliferation. The PI3K/AKT signaling pathway is considered a very important and crucial pathway that regulates the development/occurrence of MDR in cancer cells and other events such as cell cycle progression and apoptosis [65, 66]. PI3K in its active state facilitates the phosphorylation of AKT at Thr 308 and Ser 473, leading to its partial or full activation, respectively [67]. The activated AKT then partakes in controlling the inhibition of apoptosis which involves direct phosphorylation of apoptotic signal proteins or regulation of the activity of transcription factors [68].

Further, apoptosis stimulation by citrus polyphenols can lead to the migration of BCL2-associated X (BAX) on the outer membrane of mitochondria, which is a crucial key step to initiate apoptosis [69]. This sensitizes the mitochondrial permeability transition pore (MPTP) to liberate cytochrome C from the mitochondria into the cytoplasm to stimulate caspases, thus initiating cancer cell apoptosis [70]. During this process, mitochondria release apoptosis-inducing factor (AIF) into the cytoplasm, which moves to the nucleus, where it facilitates chromatin condensation, resulting in cell death [51]. This is in support of the study carried out by [71] where hesperitin in combination with cisplatin significantly caused a surge in the expression levels of PTEN and cytochrome C with a concomitant reduction in the levels of phosphorylated AKT (p-AKT) and CyclinD1. In addition, hesperetin induced a remarkable increase in AIF, BCL2-associated X, (BAX), caspases 3 and 9 and suppressed B-cell lymphoma 2 (BCL2) level [72].

Aside from the suppression of the kinase pathways for pro-apoptotic action against MDR in cancer cells, hesperidin also induces apoptosis by triggering the accumulation of reactive oxygen species (ROS) and sensitization of signal-regulating kinase 1/Jun N-terminal kinase (ASK1/JNK) pathway [73]. The high level of ROS, adenosine triphosphate (ATP) and calcium has a participatory role in the initiation of apoptosis by hesperidin in cancer cells via the sensitization of the mitochondrial pathway [74].

The arrest of cell cycle at G0/G1 phase through suppression of cyclin D1, cyclin E1 and cyclin-dependent kinase of p21, 2 (Cdk2) at the protein level with a corresponding increase in the expression levels p53 E1 is also another strategic means that hesperidin uses to change anti-apoptosis scenario in cancer cells [75].

Also, reduction of the expression of Nrf2 is known to be notoriously engaged in the modulation of drug resistance [76]. The combination of apigenin with doxorubicin during chemotherapy has been considered effective against chemoresistance treatment. Recent *in vitro* studies have also revealed that polyphenols can overcome drug resistance in cancers by inhibiting efflux pumps that extrude anticancer drugs, increasing the level of drug absorption and cell apoptosis and reducing cancer proliferation [77, 78].

4.2 In Bacteria infections

The existence of some particular structural features in flavonoids is believed to enhance their pharmacological effects, thereby describing a correlation between the flavonoid structure and its antimicrobial properties [79, 80]. For instance, a high abundance in hydroxyl groups is responsible for the increased antioxidant effects of flavonoids because of an increase in the available sites for quenching radicals and chelation of metal ions. However, high-level hydroxylation is not favorable for flavonoid lipophilicity, thus limiting their inflow across the cell membranes of the pathogen. Hence, flavonoids such as hesperetin and naringenin that are lipophilic could invade the lipid bilayer membrane, causing modifications in fluidity and accessibility of the membrane [81].

A notable number of studies have identified antimicrobial and anti-virulence as mechanisms utilized by some bactericidal flavonoids to curtail MDR through the use of specific molecular targets in microorganisms. The inhibition of the vital functions of response regulator-like transcription factor HsrA [82] such as virulence, maintaining the availability of nutrients, involvement in response to oxidative stress and cell division by these flavonoids have been described [52, 83, 84].

This can occur by blocking the interaction of HsrA with DNA at its C-terminal effector domain as observed with hesperitin [82] while other recognized molecular targets in microorganisms such as enzymes [53, 54, 85] can also be acted upon.

Another mechanism to combat the alarming increase in MDR encountered with microorganisms such as Helicobacter pylori whose infection is a risk factor for developing gastric cancer is via anti-urease activity and structure–activity relationship has proved to be beneficial. This comes to play by generation of hydrogen bonds with the amino acid residues of H. pylori urease, which is an indispensable inhibitory strategy employed by flavonoids such as quercetin. This shows the critical impact of the presence of hydroxyl groups in the structure of quercetin against *H. pylori* enzymes [86].

Inflammation has been associated with MDR that exists during infection with *H. pylori*. Cytotoxin-associated gene A (CagA) and vacuolating cytotoxin A (VacA) are considered to be very important factors during the inflammation process. Cag A is transported to host cells via the secretion system that introduces the toxin into gastric epithelial cells [87]. This triggers the activation of NF-kB with subsequent modulation of the expression of pro-inflammatory cytokines, such as IL-8, TNF-a and IL-1b [88]. Hence, there is induction of apoptosis, vacuolation and stimulation of the p38 MAPK signaling pathway in the host cell. However, nobiletin has demonstrated protection against the inflammation process linked with MDR in *H. pylori* infection by down-regulating the expression of pro-inflammatory cytokine [89]. Some of the mechanisms include a decrease in the mRNA levels of IL-8, TNF-α and IL-1b, decrease in lipid peroxidation, inhibition of vacA expression and suppression of CagA and VacA translocation to its target cells by lowering the expression of secretion components [54] and downregulation of the p38 MAPK signaling pathway [90].

The decline in the expression of the efflux pump gene hefA [91] is another mechanism citrus polyphenols utilize to reduce MDR during infections. HefA is a gene that encodes a TolC-like outer membrane channel tunnel protein and connects with translocases in the inner side of the membrane to make efflux systems partake in drug resistance [92]. The decrease in its expression increases the inhibitory capacities of antibiotics and their effectiveness in multidrug-resistant strains of bacteria [91].

Also, cell shape transformation that entails a morphological transition from spiral to coccoid forms by *H. pylori* is usually inhibited by citrus flavonoids to avoid the increase in antimicrobial resistance associated with this pathogen. This has been proven in myricetin [93].

5. Conclusion

Despite recent progress in reducing the incidence of MDR with the use of conventional drugs during chemotherapy, there is still an evident deficit in their therapeutic efficiency in inhibiting MDR. Therefore, drug resistance remains a global concern and a principal cause of cancer-related mortalities. The problem of MDR in microorganisms is one of the world's present-day challenges. However, an improved understanding of the signaling pathways involved in drug resistance would assist in paving way for more therapeutic interventions.

The application of natural products from plants is currently proving to be a viable option in targeting the various signaling pathways associated with MDR. Hence, a great deal of attention is being paid to plant-based products due to their non-toxic effects. This review demonstrated the inherent capacity of polyphenolic phytochemicals in citrus to target the various pathways utilized to perpetuate MDR and the mechanisms to circumvent it.

Conflict of interest

There is no conflict of interest.

Author details

Anne Adebukola Adeyanju
Department of Biological Sciences, KolaDaisi University, Ibadan, Nigeria

*Address all correspondence to: anne.adeyanju@koladaisiuniversity.edu.ng

References

[1] Abobatta WF. Nutritional benefits of Citrus fruits. AJBSR. 2019;**3**:4. DOI: 10.34297/AJBSR.2019.03.000681

[2] Xiao J. Dietary flavonoid aglycones and their glycosides: Which show better biological significance? Critical Reviews in Food Science and Nutrition. 2017;**57**:1874-1905. DOI: 10.1080/10408398.2015.1032400

[3] Bellavite P, Donzelli A. Hesperidin and SARS-CoV-2: New light on the healthy function of citrus fruits. Antioxidants. 2020;**9**(8):742. DOI: 10.3390/antiox9080742

[4] Frank J, Fukagawa NK, Bilia AR, Johnson EJ, Kwon O, Prakash V, et al. Terms and nomenclature used for plant-derived components in nutrition and related research: Efforts toward harmonization. Nutrition Reviews. 2020;**78**(6):451-458. DOI: 10.1093/nutrit/nuz081

[5] Cutrim CS, Cortez MSA. Review on polyphenols: Classification, beneficial effects and their application in dairy products. International Journal of Dairy Technology. 2018;**71**(3):564-578. DOI: DOI.org/10.1111/1471-0307.12515

[6] Rothwell JA, Perez-Jimenez J, Neveu V, Medina-Remón A, M'hiri N, García-Lobato P, et al. Phenol-explorer 3.0: A major update of the phenol-explorer database to incorporate data on the effects of food processing on polyphenol content. Database. 2013. DOI: 10.1093/database/bat070

[7] Alfarouk KO, Stock CM, Taylor S, Walsh M, Muddathir AK, Verduzco D, et al. Resistance to cancer chemotherapy: Failure in drug response from ADME to P-gp. Cancer Cell International. 2015;**15**(71). DOI: 10.1186/s12935-015-0221-1

[8] Prestinaci F, Pezzotti P, Pantosti A. Antimicrobial resistance: A global multifaceted phenomenon. Pathogens and Global Health. 2015;**109**(7):309-318. DOI: 10.1179/2047773215Y.0000000030

[9] Liu Z, Getzenberg ZG, Getzenberg RH, Veltri RW. The upregulation of PI3K/Akt and MAP kinase pathways is associated with resistance of microtubule-targeting drugs in prostate cancer. Journal of Cellular Biochemistry. 2015;**116**(7):1341-1349. DOI: 10.1002/jcb.25091

[10] Wu G, Qin XQ, Guo JJ, Li TY, Chen JH. AKT/ERK activation is associated with gastric cancer cell resistance to paclitaxel. International Journal of Clinical and Experimental Pathology. 2014;**7**(4):1449-1458

[11] Gao Y, Li W, Liu X, Gao F, Zhao X. Reversing effect and mechanism of soluble resistance-related calcium-binding protein on multidrug resistance in human lung cancer A549/DDP cells. Molecular Medicine Reports. 2015;**11**:2118-2124. DOI: 10.3892/MMR.2014.2936

[12] Karthikeyan S, Hoti SL. Development of fourth generation ABC inhibitors from natural products: A novel approach to overcome Cancer multidrug resistance. Anti-Cancer Agents in Medicinal Chemistry. 2015;**15**:605-615. DOI: 10.2174/1871520615666150113103439

[13] Srivalli KMR, Lakshmi PK. Overview of P-glycoprotein inhibitors: A rational outlook. Brazilian Journal of Pharmaceutical Sciences. 2012;**48**(3):353-367. DOI: 10.1590/S1984-82502012000300002

[14] Marais JPJ, Deavours B, Dixon RA, Ferreira D. The stereochemistry of flavonoids. In: Grotewold E, editor. The Science of Flavonoids. USA: Springer Science and Business Media Inc.; 2006. pp. 1-46. DOI: 10.1007/978-0-387-28822-21

[15] Iglesias-Carres L, Mas-Capdevila A, Bravo FI, Aragonès G, Muguerza B, Arola-Arnal A. Optimization of a polyphenol extraction method for sweet orange pulp (Citrus sinensis L.) to identify phenolic compounds consumed from sweet oranges. PLoS One. 2019. DOI: 10.1371/journal.pone.0211267

[16] De Rycker J, Ruxton C, Jungen M, Schweiggert R. Nutritional composition of commercially-produced 100% orange juice reveals large variability in vascular health bioactive, hesperidin. The Proceedings of the Nutrition Society. 2020;**79**. DOI: 10.1017/s0029665120002682

[17] Kay CD, Pereira-Caro G, Ludwig IA, Clifford MN, Crozier A. Anthocyanins and flavanones are more bioavailable than previously perceived: A review of recent evidence. Annual Review of Food Science and Technology. 2017;**8**:155-180. DOI: 10.1146/annurev-food-030216-025636

[18] Erlund I. Review of the flavonoids quercetin, hesperetin, and naringenin. Dietary sources, bioactivities, bioavailability, and epidemiology. Nutrition Research. 2004;**24**(10):851-874. DOI: 10.1016/j.nutres.2004.07.005

[19] Igual M, Garcia-Martinez E, Camacho MM, Martinez-Navarrete N. Jam processing and storage effects on b-carotene and flavonoids content in grapefruit. Journal of Functional Foods. 2013;**5**(2):736-737, 44. DOI: 10.1016/j.jff.2013.01.01

[20] Peterson J, Dwyer J, Beecher G, Bhagwat SA, Gebhardt SE, Haytowitz DB, et al. Flavanones in oranges, tangerines (mandarins), tangors, and tangelos: A compilation and review of the data from the analytical literature. Journal of Food Composition and Analysis. 2006a;**19**:S66-S73. DOI: 10.1016/j.jfca.2005.12.006

[21] Ioannou I, M'hiri N, Chaaban H, Boudhrioua NM, Ghoul M. Effect of the process, temperature, light and oxygen on naringin extraction and the evolution of its antioxidant activity. International Journal of Food Science & Technology. 2018;**53**(12). DOI: 10.1111/ijfs.13887

[22] Cano A, Medina A, Bermejo A. Bioactive compounds in different citrus varieties. Discrimination among cultivars. Journal of Food Composition and Analysis. 2008;**21**(5):37. DOI: 10.1016/j.jfca.2008.03.005

[23] Di Meo F, Lemaur V, Cornil J, Lazzaroni R, Durou JL, Olivier Y, et al. Free radical scavenging by natural polyphenols: Atom versus electron transfer. The Journal of Physical Chemistry. 2013;**117**(10):2082-2092. DOI: 10.1021/jp3116319

[24] Sugumar M, Sevanan M, Sekar S. Neuroprotective effect of naringenin against MPTP-induced oxidative stress. The International Journal of Neuroscience. 2019;**129**(6):534-539. DOI: 10.1080/00207454.2018.1545772

[25] Welbat JU, Naewla S, Pannangrong W, Sirichoat A, Aranarochana A, Wigmore P. Neuroprotective effects of hesperidin against methotrexate-induced changes in neurogenesis and oxidative stress in the adult rat. Biochemical Pharmacology. 2020;**178**. DOI: 10.1016/j.bcp.2020.114083

[26] Ashafaq M, Varshney L, Khan MH, Salman M, Naseem M, Wajid S, et al. Neuromodulatory effects of hesperidin in mitigating oxidative stress in streptozotocin-induced diabetes. BioMed Research International. 2014:249031. DOI: 10.1155/2014/249031

[27] Fu H, Liu L, Tong Y, Li Y, Zhang X, Gao X, et al. The antidepressant effects of hesperidin on chronic unpredictable mild stress-induced mice. European Journal of Pharmacology. 2019;**853**:236-246. DOI: 10.1016/j.ejphar.2019.03.035

[28] Zhou T, Liu L, Wang Q, Gao Y. Naringenin alleviates cognition deficits in high-fat diet-fed SAMP8 mice. Journal of Food Biochemistry. 2020;**44**(9). DOI: 10.1111/jfbc.13375

[29] Teixeira J, Chavarria D, Borges F, Wojtczak L, Wieckowski MR, Karkucinska-Wieckowska A, et al. Dietary polyphenols and mitochondrial function: Role in health and disease. Current Medicinal Chemistry. 2019;**26**(19):3376-3406. DOI: 10.2174/0929867324666170529101810

[30] Singh V. Antimicrobial resistance. In: Mendez-Vilas A, editor. Microbial Pathogens and Strategies for Combating Them: Science, Technology and Education. Vol. 1. Badajoz, Spain: Formatex Research Center; 2013. pp. 291-296

[31] Popęda M, Płuciennik ME, Bednarek AK. Proteins in cancer multidrug resistance. Advances in Hygiene and Experimental Medicine. 2014;**68**:616-632. DOI: 10.5604/17322693.1103268

[32] Tanwar J, Das S, Fatima Z, Hameed S. Multidrug resistance: An emerging crisis. Interdiscip Perspect. Infectious Diseases. 2014:541340. DOI: 10.1155/2014/541340

[33] Bennett JW, Robertson JL, Hospenthal DR, Wolf SE, Chung KK, Mende K, et al. Impact of extended-spectrum beta-lactamase producing Klebsiella pneumonia infections in severely burned patients. Journal of the American College of Surgeons. 2010;**211**(3):391-399

[34] Loeffler J, Stevens DA. Antifungal drug resistance. Clinical Infectious Diseases. 2003;**36**(1):S31-S41. DOI: 10.1086/344658]

[35] Olasehinde GI, Ojurongbe O, Adeyeba AO, Fagade OE, Valecha N, Ayanda IO, et al. *In vitro* studies on the sensitivity pattern of *plasmodium falciparum* to anti-malarial drugs and local herbal extracts. Malaria Journal. 2014;**13**:63

[36] Alekshun MN, Levy SB. Molecular mechanisms of antibacterial multidrug resistance. Cell. 2007;**128**(6):1037-1050. DOI: 10.1016/j.cell.2007.03.004

[37] He X, Li S, Kaminskyj SG. Using *aspergillus nidulans* to identify antifungal drug resistance mutations. Eukaryotic Cell. 2013;**13**(2):288-294. DOI: 10.1128/EC.00334-13]

[38] Chethana GS, Hari Venkatesh KR, Mirzaei F, Gopinath SM. Review on multidrug resistant bacteria and its implication in medical sciences. Journal of Biological and Scientific Opinion. 2013;**1**(1):32-37. DOI: 10.7897/2321-6328.01127

[39] Bansal D, Sehgal R, Chawla Y, Malla N, Mahajan RC. Multidrug resistance in amoebiasis patients. Indian Journal of Medical Research. 2006;**124**(2):189-194

[40] Coley HM. Mechanisms and strategies to overcome chemotherapy resistance in metastatic breast cancer.

Cancer Treatment Reviews. 2008;**34**:378-390. DOI: 10.1016/j.ctrv.2008.01.007

[41] Brown GA, McPherson JP, Gu L, Hedley DW, Toso R, Deuchars KL, et al. Relationship of DNA topoisomerase IIa and b expression to cytotoxicity of antineoplastic agents in human acute lymphoblastic leukemia cell lines. Cancer Research. 1996;**55**:78-82

[42] Kamath K, Wilson L, Cabral F, Jordan MA. BetaIII-tubulin induces paclitaxel resistance in association with reduced effects on microtubule dynamic instability. The Journal of Biological Chemistry. 2005;**280**(13):12902-12907. DOI: 10.1074/ jbc.M414477200

[43] Bernstein C, Bernstein H, Payne CM, Garewal H. DNA repair/ pro-apoptotic dual-role proteins in five major DNA repair pathways: Fail-safe protection against carcinogenesis. Mutation Research. 2002;**511**:145-178. DOI: 10.1016/S1383-5742(02)00009-1

[44] Simstein R, Burow M, Parker A, Weldon C, Beckman B. Apoptosis, chemoresistance, and breast cancer: Insights from the MCF-7 cell model system. Experimental Biology and Medicine. 2003;**228**(9):995-1003. DOI: 10.1177/ 153537020322800903

[45] Begicevic RR, Falasca M. ABC transporters in Cancer stem cells: Beyond Chemoresistance. International Journal of Molecular Sciences. 2017;**18**(11):2362. DOI: 10.3390/ ijms18112362

[46] Chen Z, Shi T, Zhang L, Zhu P, Deng M, Huang C, et al. Mammalian drug efflux transporters of the ATP binding cassette (ABC) family in multidrug resistance: A review of the past decade. Cancer Letters. 2016;**370**(1):153-164. DOI: 10.1016/j. canlet.2015.10.010

[47] Durmus S, Hendrikx JJ, Schinkel AH. Apical ABC transporters and cancer chemotherapeutic drug disposition. In: Adv Cancer Res. San Diego, USA: Academic Press; 2015. pp. 1-41

[48] Sun YL, Patel A, Kumar P, Chen ZS. Role of ABC transporters in cancer chemotherapy. Chinese Journal of Cancer. 2012;**31**(2):51-57. DOI: 10.5732/ cjc.011.10466

[49] Hassanzadeh P. Colorectal cancer and NF-κB signalling pathway. Gastroenterology and Hepatology from Bed to Bench. 2011;**4**(3):127-132 PMC4017424

[50] Bansal T, Jaggi M, Khar RK, Talegaonkar S. Emerging significance of flavonoids as P-glycoprotein inhibitors in Cancer chemotherapy. Journal of Pharmacy & Pharmaceutical Sciences. 2009;**12**:46-78. DOI: 10.18433/j3rc77

[51] Bano D, Prehn JHM. Apoptosis-inducing factor (AIF) in physiology and disease: The tale of a repented natural born killer. eBioMedicine. 2018;**30**:29-37. DOI: 10.1016/j.ebiom.2018.03.016

[52] Pelliciari S, Pinatel E, Vannini A, Peano C, Puccio S, De Bellis G, et al. Insight into the essential role of the Helicobacter pylori HP1043 orphan response regulator: Genome-wide identification and characterization of the DNA-binding sites. Scientific Reports. 2017;**7**:41063. DOI: 10.1038/srep41063

[53] Wu D, Kong Y, Han C, Chen J, Hu L, Jiang H, et al. D-alanine:D-alanine ligase as a new target for the flavonoids quercetin and Apigenin. International Journal of Antimicrobial Agents. 2008;**32**(5):421-426. DOI: 10.1016/j. ijantimicag.2008.06.010

[54] Yeon MJ, Lee MH, Kim DH, Yang JY, Woo HJ, Kwon HJ, et al.

Anti-inflammatory effects of Kaempferol on Helicobacter pylori-induced inflammation. Bioscience, Biotechnology, and Biochemistry. 2019;**83**(1):166-173. DOI: 10.1080/09168451.2018.1528140

[55] Kawaii S, Tomono Y, Katase E, Ogawa K, Yano M. HL-60 differentiating activity and flavonoid content of the readily extractable fraction prepared from *citrus* juices. Journal of Agricultural and Food Chemistry. 1999;**47**(1):128-135. DOI: 10.1021/jf9805101

[56] Nabekura T, Yamaki T, Kitagawa S. Effects of chemopreventive citrus phytochemicals on human P-glycoprotein and multidrug resistance protein. European Journal of Pharmacology. 2008;**600**(1-3):45-49. DOI: 10.1016/j.ejphar.2008.10.025

[57] Ambudkar S, Dey S, Hrycyna CA, Ramachandra M, Pastan I, Gottesman MM. Biochemical, cellular, and pharmacological aspects of the multidrug transporter. Annual Review of Pharmacology and Toxicology. 1999;**39**:361-398. DOI: 10.1145/ANNUREV.PHARMTOX.39.1.361

[58] Ma W, Feng S, Yao X, Yuan Z, Liu L, Xie Y. Nobiletin enhances the efficacy of chemotherapeutic agents in ABCB1 overexpression cancer cells. Scientific Reports. 2015;**5**:18789. DOI: 10.1038/srep18789

[59] Takanaga H, Ohnishi A, Yamada S, Matsuo H, Morimoto S, Shoyama Y, et al. Polymethoxylated flavones in orange juice are inhibitors of P-glycoprotein but not cytochrome P450 3A4. The Journal of Pharmacology and Experimental Therapeutics. 2000;**293**:230-236

[60] Honda Y, Ushigome F, Koyabu N, Morimoto S, Shoyama Y, Uchiumi T, et al. Effects of grapefruit juice and orange juice components on P-glycoprotein-and MRP2-mediated drug efflux. British Journal of Pharmacology. 2004;**143**(7):856-864. DOI: 10.1038/sj.bjp

[61] Febriansah R, Putri DD, Sarmoko NNA, Meiyanto E, Nugroho AE. Hesperidin as a preventive resistance agent in MCF-7 breast cancer cells line resistance to doxorubicin. Asian Pacific Journal of Tropical Biomedicine. 2014;**4**(3):228-233. DOI: 10.1016/S2221-1691(14)60236-7

[62] El-Readi MZ, Hamdan D, Farrag N, El-Shazly A, Wink M. Inhibition of P-glycoprotein activity by limonin and other secondary metabolites from Citrus species in human colon and leukaemia cell lines. European Journal of Pharmacology. 2010;**626**(2-3):139-145. DOI: 10.1016/j.ejphar.2009.09.040

[63] Ferreira de Oliveira JMP, Santos C, Fernandes E. Therapeutic potential of hesperidin and its aglycone hesperetin: Cell cycle regulation and apoptosis induction in cancer models. Phytomedicine. 2019;**73**. DOI: 10.1016/j.phymed.2019.152887

[64] Li WX, Chen X, Yang Y, Huang HM, Li HD, Huang C, et al. Hesperitin derivative-11 suppress hepatic stellate cell activation and proliferation by targeting PTEN/AKT pathway. Toxicology. 2017;**381**:75-86. DOI: 10.1016/j.tox.2016.11.004

[65] Warfel NA, Kraft AS. PIM kinase (and Akt) biology and signaling in tumors. Pharmacology & Therapeutics. 2015;**151**:41-49. DOI: 10.1016/j.pharmthera.2015.03.001

[66] Spangle JM, Roberts TM, Zhao JJ. The emerging role of PI3K/AKT-mediated epigenetic regulation in cancer. Biochimica Et Biophysica Acta. Reviews on Cancer. 2017;**1868**(1):123-131. DOI: 10.1016/j.bbcan.2017.03.002

[67] Abraham J. PI3K/AKT/mTOR pathway inhibitors: The ideal combination partners for breast cancer therapies? Expert Review of Anticancer Therapy. 2015;**15**(1):51-68. DOI: 10.1586/14737140.2015.961429

[68] Shaw RJ, Cantley LC. Ras, PI(3) K and mTOR signalling controls tumour cell growth. Nature. 2006;**441**(7092):424-430. DOI: 10.1038/nature04869)

[69] Pena-Blanco A, Garcia-Saez AJ. Bax, Bak and beyond –mitochondrial performance in apoptosis. Federation of European Biochemical Societies Journal. 2018;**285**(3):416-431. DOI: 10.1111/febs.14186

[70] Cosentino K, Garcia-Saez AJ. Bax and Bak pores: Are we closing the circle? Trends in Cell Biology. 2017;**27**(4):266-275. DOI: 10.1016/j.tcb.2016.11.004)

[71] He P, Ma J, Liu Y, Deng H, Dong W. Hesperetin promotes cisplatin-induced apoptosis of gastric Cancer *In vitro* and *In vivo* by upregulating *PTEN* expression. Frontiers in Pharmacology. 2020;**27**(11):1326. DOI: 10.3389/fphar.2020.01326

[72] Siddiqi A, Hasan SK, Nafees S, Rashid S, Saidullah B, Sultana S. Chemopreventive efficacy of hesperidin against chemically-induced nephrotoxicity and renal carcinogenesis via amelioration of oxidative stress and modulation of multiple molecular pathways. Experimental and Molecular Pathology. 2015;**99**:641-653. DOI: 10.1016/j.yexmp.2015.11.012

[73] Palit S, Kar S, Sharma G, Das PK. Hesperetin induces apoptosis in breast carcinoma by triggering the accumulation of ROS and activation of ASK1/JNK pathway. Journal of Cellular Physiology. 2015;**230**:1729-1739. DOI: 10.1002/jcp.24818

[74] Zhang J, Song J, Wu D, Wang J, Dong W. Hesperetin induces the apoptosis of hepatocellular carcinoma cells via mitochondrial pathway mediated by the increased intracellular reactive oxygen species, ATP and calcium. Medical Oncology. 2015;**32**:101. DOI: 10.1007/s12032-015-0516-z

[75] Wang Y, Yu H, Zhang J, Gao J, Ge X, Lou G. Hesperidin inhibits Hela cell proliferation through apoptosis mediated by endoplasmic reticulum stress pathways and cell cycle arrest. BMC Cancer. 2015;**15**:682. DOI: DOI.org/10.1186/s12885-015-1706-y

[76] Ji L, Li H, Gao P, Shang G, Zhang DD, Zhang N, et al. Nrf2 pathway regulates multidrug-resistance-associated protein 1 in small cell lung Cancer. PLoS One. 2013;**8**(5). DOI: 10.1371/jounal.pone.0063404

[77] Dana PM, Sadoughi F, Asemi Z, Yousefi B. The role of polyphenols in overcoming cancer drug resistance: A comprehensive review. Cellular & Molecular Biology Letters. 2022;**3**(27):1. DOI: 10.1186/s11658-021-00301-9

[78] Gu HF, Mao XY, Du M. Prevention of breast cancer by dietary polyphenols-role of cancer stem cells. Critical Reviews in Food Science and Nutrition. 2020;**60**(5):810-825. DOI: 10.1080/10408398.2018.1551778

[79] Farhadi F, Khameneh B, Iranshahi M, Iranshahy M. Antibacterial activity of flavonoids and their structure-activity relationship: An update review. Phytotherapy Research. 2019;**33**(1):13-40. DOI: 10.1002/ptr.6208

[80] Górniak I, Bartoszewski R, Króliczewski J. Comprehensive review

of antimicrobial activities of plant flavonoids. Phytochemistry Reviews. 2019;**18**(1):241-272. DOI: 10.1007/s11101-018-9591-z

[81] Tarahovsky YS, Kim YA, Yagolnik EA, Muzafarov EN. Flavonoid-membrane interactions: Involvement of flavonoid-metal complexes in raft Signaling. Biochimica et Biophysica Acta. 2014;**1838**(5):1235-1246. DOI: 10.1016/j.bbamem.2014.01.021

[82] González A, Salillas S, Velázquez-Campoy A, Espinosa Angarica V, Fillat MF, Sancho J, et al. Identifying potential novel drugs against Helicobacter pylori by targeting the essential response regulator HsrA. Scientific Reports. 2019;**9: (1)**:11294. DOI: 10.1038/s41598-019-47746-9)

[83] Olekhnovich IN, Vitko S, Chertihin O, Hontecillas R, Viladomiu M, Bassaganya-Riera J, et al. Mutations to essential orphan response regulator HP1043 of Helicobacter pylori result in growth-stage regulatory defects. Infection and Immunity. 2013;**81**(5):1439-1449. DOI: 10.1128/IAI.01193-12

[84] Olekhnovich IN, Vitko S, Valliere M, Hoffman PS. Response to metronidazole and oxidative stress is mediated through homeostatic regulator HsrA (HP1043) in Helicobacter pylori. Journal of Bacteriology. 2014;**196**(4):729-739. DOI: 10.1128/JB.01047-13

[85] Moon SH, Lee JH, Kim KT, Park YS, Nah SY, Ahn DU, et al. Antimicrobial effect of 7-O-Butylnaringenin, a novel flavonoid, and various natural flavonoids against Helicobacter pylori strains. International Journal of Environmental Research and Public Health. 2013;**10**(11):5459-5469. DOI: 10.3390/ijerph10115459

[86] Yu XD, Zheng RB, Xie JH, Su JY, Huang XQ, Wang YH, et al. Biological evaluation and molecular docking of Baicalin and Scutellarin as Helicobacter pylori urease inhibitors. Journal of Ethnopharmacology. 2015;**162**:69-78. DOI: 10.1016/j.jep.2014.12.041

[87] Backert S, Haas R, Gerhard M, Naumann M. The Helicobacter pylori type IV secretion system encoded by the cag Pathogenicity Island: Architecture, function, and Signaling. Current Topics in Microbiology and Immunology. 2017;**413**:187-220. DOI: 10.1007/978-3-319-75241-9_8

[88] Lamb A, Chen LF. Role of the Helicobacter pylori-induced inflammatory response in the development of gastric Cancer. Journal of Cellular Biochemistry. 2013;**114**(3):491-497. DOI: 10.1002/jcb.24389

[89] Ouyang Y, Li L, Ling P. Nobiletin inhibits Helicobacterium pylori infection-induced gastric carcinogenic Signaling by blocking inflammation, apoptosis, and mitogen-activated protein kinase events in gastric epithelial 1 cells. Journal of Environmental Pathology, Toxicology and Oncology. 2020;**39**(1):77-88. DOI: 10.1615/JEnvironPatholToxicolOncol.2020031272

[90] Zhang S, Huang J, Xie X, He Y, Mo F, Lu Z. Quercetin from Polygonum capitatum protects against gastric inflammation and apoptosis associated with Helicobacter pylori infection by affecting the levels of P38mapk, BCL-2 and BAX. Molecules. 2017;**22**(5):744. DOI: 10.3390/molecules22050744)

[91] Huang YQ, Huang GR, Wu MH, Tang HY, Huang ZS, Zhou XH, et al. Inhibitory effects of Emodin, Baicalin, Schizandrin and Berberine on hefA gene: Treatment of Helicobacter pylori-induced multidrug resistance.

World Journal of Gastroenterology. 2015;**21**(14):4225-4231. DOI: 10.3748/wjg.v21.i14.4225), 10.3748/wjg.v21.i14.4225)

[92] Liu ZQ, Zheng PY, Yang PC. Efflux pump gene hefA of Helicobacter pylori plays an important role in multidrug resistance. World Journal of Gastroenterology. 2008;**14**(33):5217-5222. DOI: 10.3748/wjg.14.5217

[93] Krzyzek P, Migdal P, Paluch E, Karwanska M, Wieliczko A, Gosciniak G. Myricetin as an Antivirulence compound interfering with a morphological transformation into coccoid forms and potentiating activity of antibiotics against Helicobacter pylori. International Journal of Molecular Sciences. 2021;**22**(5):2695. DOI: 10.3390/ijms22052695

Chapter 11

Citrus Essential Oils and Nanosystems towards Skin Delivery

Júlia Scherer Santos and Mateus Pereira Gonzatto

Abstract

Essentials oils from citrus have anti-inflammatory and antioxidant activity. Furthermore, terpenes are their main phytochemicals, namely limonene is the most important one. As terpenes are permeation promoters, they have been used to improve transdermal delivery of drugs. In addition, a proper oil source is a key factor to obtain desired phytochemicals. Recently, polymeric nanoparticles, solid lipid nanoparticles, nanostructured lipid carriers, nanoemulsions, liposomes and elastic liposomes as carriers of citrus essential oils or citrus terpenes have been developed to achieve more effective formulations. In this chapter, the most recent publications on nanocarriers containing citrus oils or citrus terpenes were addressed. In that regard, citrus oil or terpenes loaded in nanotechnological systems improve drugs skin permeation. Besides, terpenes loaded in nanoparticles also increase transdermal delivery of drugs. As essential oils and their respective terpenes are volatile compound and prone to oxidation, its encapsulations reduce oxidation and volatility. Hence, an improved antioxidant activity can be obtained. Therefore, nanoformulations of citrus oils or citrus terpenes are potential approaches to skin topical and transdermal delivery.

Keywords: terpenes, nanocarriers, limonene, citrus oils, transdermal, skin delivery

1. Introduction

Citrus essential oils are widely used in perfumery due to their very pleasant aroma. Nevertheless, because of the high volatility, the aroma is quickly lost [1], hindering its cosmetic application. Regarding medicinal properties, antioxidant, antifungal and anticancer ones are reported [2]. Moreover, citrus oil would be interesting to inflammatory skin diseases [3] and to disorders generating oxygen-generating species [4, 5]. Then, citrus oils could be used as adjuvants in the treatment of several skin disorders.

In their composition, essential oils have terpenes [6] which are employed in pharmaceutics formulations to provide a transdermal release due to its ability to increase skin permeation [7]. Despite its limited use, transdermal route may be used to overcome the drawbacks of the injectable routes or as an alternative to the oral route allowing a systemic drug delivery through the skin [8]. Additionally, as several factors affect oil chemical composition [9], it is crucial to select an adequate oil source to obtain a suitable phytochemical composition. Hence, in the development of therapeutic products plant-ingredient based, it is required to associate strategies of a correct plant cultivation together with an adequate pharmaceutical study [3, 4].

Nanoparticulate systems are promising approaches to dermatological [10] and systemic-diseases treatment [11]. Accordingly, this chapter addresses nanoformulations containing citrus terpenes or citrus essential oils targeting skin delivery. Initially, aspects related to chemical composition will be briefly displayed. Lastly, the most recent contributions of nanocarriers containing terpenes or essential oils from citrus for skin or transdermal delivery will be shown.

2. Citrus oil composition

The essential oils composition varies among different species and cultivars, rootstocks, structure (of leaf, flower or fruit), according to the extraction method, edaphoclimatic conditions and agronomic management employed at harvest time. In general, the constituents of the essential oil present in the peel of citrus fruits are monoterpenes, followed by sesquiterpenes and phenylpropanoids [12, 13].

Oranges fruit juice is the main source of citrus oils. Besides, the peel of young green fruits removed during the manual thinning of mandarin fruits is also an origin of essential oil. Moreover, leaves and flower buds are other sources of oils, by steam distillation or hydrodistillation extraction methods. Oils obtained from flower buds are known as petitgrain (*petit grain* in French means "small grain") as an allusive name to the leaves and buds of the bitter orange (*Citrus aurantium*). However, petitgrain oil can be extracted from leaves of any citrus species [12, 14, 15].

In most citrus oils, the major terpene is d-limonene, accounting for a percentage of up to 98%. Other compounds of importance, such as γ-terpinene, geraniol, citral, valencene, α-pinene, sabinene, myrcene and linalool, are present in variable amounts [12]. Essential oils extracted from the peel of sweet oranges (*Citrus sinensis*), grapefruit (*Citrus paradisi*), pummelo (*Citrus grandis*), clementines (*Citrus clementina*), satsumas (*Citrus unshiu*) and tangors (hybrids of orange with tangerines) have a limonene content higher than 90% [16–19]. On the other hand, the contents of limonene and linalyl acetate are similar (~35%) in bergamot oil (*Citrus bergamia*) [18]. Elseways, limonene content ranges from around 70 to 75% in Cai and Montenegrina mandarin (*Citrus deliciosa*) [20].

With respect to essential oils extracted from citrus leaves (petitgrain), there is a greater variation between species. A greater amount of methyl-N-methylanthranilate is found in *Citrus deliciosa* species. Contrarily, *Citrus reticulata*, *Citrus reshni* and tangors seem to accumulate higher contents of linalool and sabinene. Diversely, in *Citrus sunki*, beta-pinene is the most important terpene [14, 15].

3. Nanocarriers bearing citrus oils or citrus terpenes

Nanopharmaceuticals either contain terpenes or essential oils. The use of isolated terpenes is desirable at times [11–14] instead of essential oil. Concerning citrus oils, they are related to skin irritation and low stability due to their volatility. To overcome these limitations, nanoencapsulation of these oils has been reported [21]. Thereby, the loading of essential oils in microemulsions allows to reduce adverse effects [21], and the encapsulation of citrus oil in ethosomes reduces oil volatilization [22].

Nanodrug delivery systems include polymeric nanoparticles, liposomes, ethosomes, nanoemulsions, solid lipid nanoparticles, nanostructured lipid carriers and microemulsions [23]. **Figure 1** shows the most important nanometric carriers bearing citrus oils or citrus terpenes. Polymeric nanoparticles are composed of polymers and surfactants in which different drugs can be encapsulated. Liposomes are vesicular systems containing phospholipids [26]. Other vesicular carriers have been developed over the years including ethosomes [11, 27], invasomes [28, 29], transfersomes [11] and bilosomes [30], and they all have other ingredients than phospholipids [25, 26]. For instance, in bilosomes, there are bile salts [30].

Nanoemulsions contain an oily core and surfactants in their composition. Solid lipid nanoparticles (SLN) and nanostructured lipid carriers (NLC) are both formed of lipids where SLN and NLC contain, respectively, solid lipids and a mixture of solid and liquid lipids (**Figure 1**) [26]. Elseways, microemulsions are similar to nanoemulsions as they have an oil phase, an aqueous phase, and surfactants (**Figure 1**) [31, 32]. Nonetheless, microemulsions are thermodynamically stable systems [33], unlike nanoemulsions [32]. These drug delivery systems are mainly employed to entrap oily substances and therefore are suitable for essential oils [10, 21, 22, 34, 35].

Previously to nanoencapsulation, essential oils are extracted [4] or at times commercial essential oils can be used [10, 22, 35]. Either terpenes or essential oils are employed in nanostructures as a therapeutic ingredient [4, 21, 34] or as skin permeation enhancers [36–40]. About terpenes, limonene provides a drug release into the skin [36] or a drug transdermal release [39–41].

Limonene is frequently used in liposomes to obtain invasomes. Invasomes are vesicular systems composed of terpenes and alcohol [24] which favour the skin penetration/permeation of drugs [42]. Apart from that, when topical application of drugs is mentioned both permeation and penetration may occur. Permeation is the passage of drugs from one skin layer to another skin layer (e.g. from the epidermis to the dermis). Oppositely, penetration is when a drug reaches one skin layer (e.g. the stratum corneum). Skin penetration of drugs always precedes permeation. Once skin permeation is reached, drugs may reach systemic circulation [43].

In general, nanocarriers show a biphasic pattern with an initial burst drug release followed by a sustained release. The initial release is due to the outermost drug

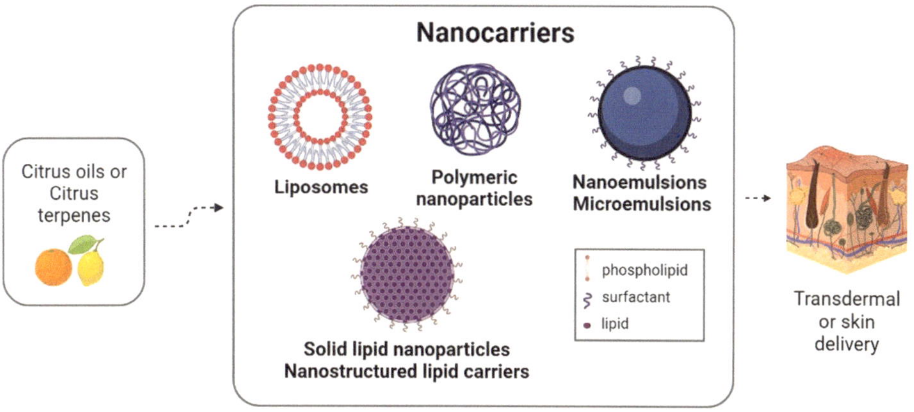

Figure 1.
Citrus essential oils or citrus terpenes loaded in nanocarriers targeting transdermal or skin delivery. Source: [24, 25]. Created in BioRender.

location in the nanocarrier. It is expected that at least some part of the drug will be adsorbed to the nanostructure, favouring its burst release. Drugs located more internally in the nanocarrier will show a gradual release providing the prolonged release effect. The combination of initial burst release and prolonged release is interesting because it allows a pharmacological activity just after skin application along to a prolonged effect [30, 39], even for several hours [30, 35, 38].

Furthermore, ingredients of nanometric carriers must be carefully selected considering the concentrations of each ingredient. From an optimized formulation, subsequent efficacy and safety studies can be conducted [30, 38, 44]. An optimized formulation can be determined by physicochemical assays including particle size, zeta potential and encapsulation efficiency [10, 11, 30, 35, 37, 45]. Also, optimization also implies evaluation of formulation stability over a period of weeks or months at different temperatures [7–49].

Concerning nanometric systems intended for skin application, both a localized effect on the skin [11, 31, 34, 36, 37] or a systemic effect can occur [29, 41, 45]. A topical effect is desirable to skin diseases treatment [11, 31, 36, 37] while a systemic effect is desirable to treat a more serious disease whose source is in another organ (i.e. not the skin) [30, 41, 45]. For topical delivery, skin penetration and/or permeation is desired [36]. Nevertheless, for systemic delivery, permeation is mandatory as it precedes drug release into the circulatory stream [30, 39]. Additionally, in view of the trend of using natural compounds, nanoparticles are also applied to increase skin penetration of polyphenols [36].

Since the discovery of new drugs is very expensive and very time consuming [50], the use of penetration enhancers improves biopharmaceutical characteristics of drugs (e.g. low aqueous solubility) [38, 44]. Hence, combination of physical and chemical permeation methods is proposed as a strategy to increase skin permeability of drugs [39, 45].

3.1 Citrus oils loaded into nanocarriers

Most citrus oils were encapsulated in lipid carriers, as shown in **Table 1**. Also, spanlastics, ethosomes and transfersomes were employed as carriers due to their ability to increase skin penetration. These carriers are entitled as elastic liposomes, variations of conventional liposomes [51]. Since conventional liposomes (**Figure 1**) generally provide skin penetration only at the outermost skin layer, elastic liposomes are used to improve skin penetration into the innermost layers of the skin [52]. In this sense, elastic liposomes are composed of penetration enhancers such as ethanol and surfactants [7]. Ethosomes and transfersomes are vesicular systems containing ethanol and surfactants, respectively [11]. On the other hand, spanlastics are elastic liposomes bearing the surfactants span 60 and polysorbate 80 [51].

Recently, lipid nanocapsules were proposed as blending the features of polymeric nanoparticles and of liposomes. In its composition, there is lecithin, an ingredient commonly used for liposomes. This carrier resembles polymeric nanoparticles as it contains an oil phase and an aqueous phase containing hydrophilic surfactants [53]. In the case of lipid nanocapsules, polyethylene glycol derivatives are employed as hydrophilic surfactants. This new carrier may be more attractive than traditional polymeric nanoparticles in increasing the skin permeation [34].

Essential oils content varies according to the type of nanocarrier (**Table 1**). Most studies employed low essential oils content [10, 22, 35]. However, for lipid

Nanocarrier	Essential oil	EO content (%)	Drug	Main outcome	Reference
Spanlastics	Bergamot oil	1	—	Lower time to skin repigmentation	[10]
Nanostructured lipid carriers	Bergamot oil	6.9	—	Lower to skin repigmentation	[35]
Ethosomes	Orange oil	7	—	Higher aroma durability	[22]
Lipid nanocapsules	Orange oil	40	—	Higher inhibition of fungal growth	[34]
Microemulsion	Lemon oil	30	Resveratrol	Higher skin permeation and higher antioxidant activity	[31]
Nanoemulsion	Bergamot oil	14.5	Luteolin	Higher skin permeation	[47]

EO: essential oil.

Table 1.
Nanocarriers bearing citrus essential oil.

nanocapsules [34] and microemulsions [31], a higher concentration was used. Although a high oil concentration may provide a formulation with a lower physical stability [22], a higher oil content may be needed to ensure nanocarrier formation (e.g. lipid nanocapsules) [34]. Therefore, it is fundamental to ensure a suitable oil content as it may influence nanosystems performance.

As for physicochemical characterization, essential oil content influenced particle size. The higher the oil content the higher particle size [22]. Diversely, the lower the concentration of surfactants the higher the particle size [10, 38, 41], which is related to the stabilizing effect of surfactants. A higher surfactant concentration improves stability, reduces particle size, and prevents coalescence [54]. Zeta potential, another important assay, also is also indicative of stability [38]. It provides information about the surface electrical charge, and it is affected by anionic and cationic surfactants. Cationic surfactant will impart a positive zeta potential, while anionic surfactant will provide a negative zeta potential [47]. Regards to encapsulation efficiency, it only applies when a drug is entrapped inside a nanocarrier [31, 47]. A high encapsulation efficiency means that a high percentage of drug is encapsulated. In that regard, a high encapsulation efficiency was obtained for luteolin-loaded in bergamot oil nanoemulsions, a desirable effect as luteolin is protected from degradation [47].

Another strategy is the development of cationic carriers to increase skin permeation. Luteolin-loaded in cationic nanoemulsions had a better skin permeation regarding luteolin-loaded anionic nanoemulsions. Since there is electrostatic interaction between the skin and cationic nanocarrier, the skin adhesion is favoured. Consequently, there is a greater permeation of luteolin. Nonetheless, luteolin loaded in anionic nanoemulsions had a higher skin permeation than non-nanotechnological luteolin suspensions. Although the use of cationic carrier usually provides the best skin penetration, anionic carrier are also potential carriers are they may have a better performance than non-nanotechnological formulations [47].

Citrus essential oils are related to phototoxic reactions due to the presence of furocoumarins [55]. Nonetheless, for vitiligo, this phototoxic reaction can be desirable to stimulate skin repigmentation [10]. In this sense, bergamot oil-loaded lipid carriers had an increased skin repigmentation in vitiligo patients [10, 35] due to the increased skin permeation of bergamot oil loaded into nanocarriers [35]. In view of vitiligo treatment limitations and treatment-related side effects [10], vesicular lipid carriers are a promising treatment for vitiligo patients.

Further, orange oil entrapped into nanocarriers provided a greater antifungal activity [34], probably due to oil protection from volatilization. Likewise, orange oil encapsulation also promoted an increase in aroma durability, a desirable effect in perfumery [22]. As citrus essential oils are top notes [1], its encapsulation may be an interesting alternative in the development of perfumes with greater durability.

Newly, essential oils from *Citrus aurantifolia and C. reticulata* were loaded in microemulsions. Citrus oils loaded in microemulsions were less irritating to the skin than undiluted oil, due to oil-controlled release provided by microemulsions [4]. Although limonene may cause change in epidermis structure to promote skin penetration [56], citrus essential oils-loaded microemulsions were less irritant than undiluted oils [4].

3.2 Citrus terpenes loaded into nanocarriers

Limonene is the most employed terpene in drug delivery systems [29, 37, 39, 44] while linalool is least frequent [48, 57]. Encapsulation of linalool improves its stability [48, 57] and consequently increases its antioxidant activity since the terpene itself has radical-scavenging ability [58]. The higher radical-scavenging ability is important to ensure a more effective skin treatment as oxidative stress can worse dermatological diseases [59, 60].

Table 2 summarizes the main outcomes of terpenes entrapped into nanocarriers. As the log P of limonene is greater than of linalool, the former is a more suitable ingredient for lipophilic drugs [64]. In fact, a higher encapsulation efficiency is reported to nanocarriers bearing limonene and lipophilic drugs [29, 39, 44]. As to terpenes content, it is variable, between 0.47% [46] and 5% [49]. As can be noted by comparison of **Tables 1** and **2**, a lower terpenes content is employed regarding essential oils content.

Moreover, particle size [24, 39, 44] and drug release [37, 45] are affected by limonene content. Particle size is a key feature in skin permeation. In general, carriers with smaller particle sizes are more efficient in promoting permeation [65]. As for drug release, a higher limonene content caused a lower drug release and possibly generating a prolonged release [37]. Limonene is then the most suitable terpene to provide a greater release of lipophilic drugs due to its lower boiling point [37].

Several studies have shown a better performance of limonene entrapped into lipid carriers [28, 37, 39, 63] for transdermal drug release [37, 39, 44] or drug topical release [11, 36]. Despite nanosystems itselves also having the ability to promote skin penetration [66], nanoencapsulation of limonene [40, 63] further increases the penetration of drug into skin [40, 63]. In this sense, limonene-pegylated liposomes improved skin permeation of galantamine in relation to pegylated liposomes [40].

About carriers containing terpenes, invasomes [29, 39], bilosomes [30], microemulsions [31] and nanoemulsions [36] are described. Invasomes provide a greater drug skin permeation [39, 63] than conventional liposomes [63] and other nanocarriers [29]. Solid lipid nanoparticles and nanostructured lipid carriers promote an occlusive effect which causes an increase drug penetration [65]. In turn, lornoxicam-loaded in bilosomes had a

Nanocarrier	Terpene	Terpene content (%)	Drug	Main outcome	Reference
Liposomes	LI/LIN	3	—	Higher antioxidant activity	[58]
	LI	2	Galantamine	Higher skin permeation	[56]
Bilosomes	LI	0.47	Lornoxicam	Improvement in anti-inflammatory activity	[30]
Polymeric nanoparticles	LIN	0.5	—	Lower loss of linalool	[57]
Solid lipid nanoparticles	LIN	1	—	Stability improvement	[48]
Nanoemulsions	LI	1	Resveratrol	Higher content in skin	[36]
		0.05	Astaxanthin	Higher stability/ higher skin content	[61]
Nanoemulsions	LI	5	Ibuprofen	Higher skin permeation	[62]
Microemulsions		3		Higher *in vitro* release	
Transfersomes	LI	0.9	Cyclosporin A	Higher skin penetration	[11]
Invasomes	LI	1.5	Agomelatine	Greater permeation for skin pre-treated with ultrasound	[39]
		1	Asenapine	Higher skin permeation	[29]
		1	Phenylethyl resorcinol	Lower tyrosinase activity	[63]
		0.5 and 1.0	Tizanidine	Greater permeation for skin pre-treated with microneedling	[45]
		1.5	Avanafil	Higher bioavailability	[28]
		1.5	Dapsone	Higher bioavailability	[37]

**LI: limonene, LIN: linalool.*

Table 2.
Nanocarriers bearing citrus terpenes.

transdermal release resulting in a better anti-inflammatory effect. As bilosomes have bile salts in their composition, a higher skin permeability can be achieved [30]. Oppositely, nanoemulsions and microemulsions are used to increase solubility of low-solubility drugs which results in an enhanced topical delivery [31, 36].

Moreover, nanoemulsions [36, 61, 62], transfersomes [11], invasomes [63] and microemulsions [61] are also employed for purposes of skin topical effect. Phenylethyl resorcinol loaded in elastic liposomes had a lower tyrosinase activity along with a higher skin deposition [63]. Transfersomes improved skin deposition of cyclosporin and would be an interesting topical treatment for psoriasis [11]. Resveratrol nanoemulsions [36] and astaxanthin self-nanoemulsifying systems [61] had a higher photostability [36] and a higher antioxidant ability and then can be used in the treatment of different skin conditions [61]. Additionally, due to limonene, there was an improved deposition in skin layers to astaxanthin [61] and resveratrol [36] loaded in delivery systems.

Concerning systemic delivery, nanocarriers can improve drug bioavailability [28, 44], the amount of drug that reaches the circulatory stream and is directly related to the pharmacological effect [67]. Avanafil loaded-invasomes transdermal films [28], raloxifene in transfersomes gels [44] and dapsone-loaded invasomes [37] had a higher bioavailability than non-nanotechnological formulations. The increased bioavailability suggests a possible use of skin as an alternative route of drug administration, notably for drugs with low oral bioavailability [29, 44].

Other report showed different performance of ibuprofen-microemulsions and ibuprofen-nanoemulsions for topical delivery. Microemulsions provided a greater drug release while nanoemulsions were the most effective regarding skin penetration. The better skin penetration for nanoemulsions probably is due to its higher terpene content, and the higher drug release for microemulsions is caused by its higher interfacial area. Aside from this, nanoemulsions also had a higher cell viability. Hence, nanoemulsions were the most appropriate carriers for cutaneous delivery [62].

Recently, a skin pre-treatment with ultrasound increased cutaneous deposition of agomelatine-loaded invasomes [39]. Similarly, microneedling pre-treatment increased the permeation of tizanidine-loaded invasomes [45]. Therefore, a combination of physical and chemical methods of skin penetration has been designed to improve transdermal delivery [68].

4. Conclusion

Citrus plants are cultivated on all continents, and derived products are used worldwide. Although citrus oils are being widely used as fragrances, they have limited use in dermatological treatments. A trend over plant ingredients loaded in nanocarriers for topical products is seen due to the several benefits provided by vegetal phytocompounds. In addition, further development of products bearing plant ingredients can be increased by its nanoencapsulation in drug delivery systems.

A few studies report clinical trials of nanoformulations based on essential oils or on citrus terpenes. It is expected that more safety and efficacy on humans will be conducted so that new nanomedicines become available. Hence, the quality of life of patients having diseases whose treatment is side effects associated may be improved. Likewise, the improvement of transdermal drug release by combination of several skin penetration enhancers may be a potential approach to develop less invasive pharmacological treatments.

Author details

Júlia Scherer Santos[1]* and Mateus Pereira Gonzatto[2]

1 Faculty of Pharmacy, Department of Pharmaceutical Sciences, Federal University of Juiz de Fora, Juiz de Fora, Brazil

2 Department of Agronomy, Federal University of Viçosa, Viçosa, Brazil

*Address all correspondence to: julia_scherer_santos@hotmail.com

References

[1] Sharmeen JB, Mahomoodally FM, Zengin G, Maggi F. Essential oils as natural sources of fragrance compounds for cosmetics and cosmeceuticals. Molecules. 2021;**26**(3):2-24. DOI: 10.3390/molecules26030666

[2] Raut JS, Karuppayil SM. A status review on the medicinal properties of essential oils. Industrial Crops and Products. 2014;**62**:250-264. DOI: 10.1016/j.indcrop.2014.05.055

[3] Maurya AK, Mohanty S, Pal A, Chanotiya CS, Bawankule DU. The essential oil from Citrus limetta Risso peels alleviates skin inflammation: In-vitro and in-vivo study. Journal of Ethnopharmacology. 2018;**212**:86-94. DOI: 10.1016/j.jep.2017.10.018

[4] Prommaban A, Chaiyana W. Microemulsion of essential oils from citrus peels and leaves with anti-aging, whitening, and irritation reducing capacity. Journal of Drug Delivery Science Technology. 2022;**69**:103188. DOI: 10.1016/j.jddst.2022.103188

[5] Frassinetti S, Caltavuturo L, Cini M, Croce Della CM, Maserti BE. Antibacterial and antioxidant activity of essential oils from citrus spp. Journal of Essential Oil Research. 2011;**23**(1):27-31. DOI: 10.1080/10412905.2011.9700427

[6] Noriega P. Terpenes in essential oils: Bioactivity and applications. In: Perveen S, Al-Taweel AM, editors. Terpenes and Terpenoids. London: IntechOpen; 2021. pp. 1-13. DOI: 10.5772/intechopen.93792

[7] Williams AC, Barry BW. Penetration enhancers. Advanced Drug Delivery Reviews. 2004;**56**(5):603-618. DOI: 10.1016/j.addr.2003.10.025

[8] Prausnitz MR, Langer R. Transdermal drug delivery. Nature Biotechnology. 2008;**26**(11):1261-1268. DOI: 10.1038/nbt.1504

[9] Figueiredo AC, Barroso JG, Pedro LG, Scheffer JJC. Factors affecting secondary metabolite production in plants: Volatile components and essential oils. Flavour and Fragrance Journal. 2008;**23**(4):213-226. DOI: 10.1002/ffj.1875

[10] Shaaban M, Nasr M, Tawfik AA, Fadel M, Sammour O. Novel bergamot oil nanospanlastics combined with PUVB therapy as a clinically translatable approach for vitiligo treatment. Drug Delivery and Translational Research. 2019;**9**(6):1106-1116. DOI: 10.1007/s13346-019-00653-y

[11] Carreras JJ, Tapia-Ramirez WE, Sala A, Guillot AJ, Garrigues TM, Melero A. Ultraflexible lipid vesicles allow topical absorption of cyclosporin a. Drug Delivery and Translational Research. 2020;**10**(2):486-497. DOI: 10.1007/s13346-019-00693-4

[12] EFROM CFS, SOUZA PVD de. Citricultura no Rio Grande do Sul. Porto Alegre: SEAPI; 2018. p. 289

[13] Verzera A, Trozzi A, Gazea F, Cicciarello G, Cotroneo A. Effects of rootstock on the composition of bergamot (*Citrus bergamia* Risso et Poiteau) essential Oil. Journal of Agricultural and Food Chemistry. 2003;**51**(1):206-210. DOI: 10.1021/jf0206872

[14] Sulzbach M, da Silva MA, Gonzatto MP, Marques MMO, Böettcher GN, Silvestre WP, et al. Effect of distillation methods on the leaf essential oil of some *Citrus* cultivars.

Journal of Essential Oil Research. 2021;**33**(5):452-463. DOI: 10.1080/10412905.2021.1936666

[15] Silvestre WP, Sachett FH, Agostini F, Boettcher GN, Sulzbach M, Gonzatto MP, et al. Chemical composition of petitgrain (leaf) essential oil of different *Citrus* rootstocks and scion cultivars. Journal of Essential Oil Research. 2020;**32**(5):394-406. DOI: 10.1080/10412905.2020.1787886

[16] Perini JF, Silvestre WP, Agostini F, Toss D, Pauletti GF. Fractioning of orange (*Citrus sinensis* L.) essential oil using vacuum fractional distillation. Separation Science and Technology. 2017;**52**(8):1397-1403. DOI: 10.1080/01496395.2017.1290108

[17] Njoroge SM, Koaze H, Karanja PN, Sawamura M. Volatile constituents of Redblush grapefruit (*Citrus paradisi*) and Pummelo (*Citrus grandis*) Peel essential oils from Kenya. Journal of Agricultural and Food Chemistry. 2005;**53**(25):9790-9794. DOI: 10.1021/jf051373s

[18] Verzera A, Trozzi A, Cotroneo A, Lorenzo D, Dellacassa E. Composition of Nova and Satsuma Mandarin oils. Journal of Agricultural and Food Chemistry. 2000;**48**(7):2903-2909. DOI: 10.1021/jf990734z

[19] Ruberto G, Renda A, Piattelli M, Rapisarda P, Starrantino A. Essential Oil of two new pigmented Citrus hybrids, *Citrus Clementina* × *Citrus sinensis*. Journal of Agricultural and Food Chemistry. 1997;**45**(2):467-471. DOI: 10.1021/jf960109j

[20] Frizzo CD, Lorenzo D, Dellacassa E. Composition and seasonal variation of the essential oils from two mandarin cultivars of southern Brazil. Journal of Agricultural and Food Chemistry. 2004;**52**(10):3036-3041. DOI: 10.1021/jf030685x

[21] Mehanna MM, Abla KK, Elmaradny HA. Tailored limonene-based nanosized microemulsion: Formulation, physicochemical characterization and in vivo skin irritation assessment. Advanced Pharmaceutical Bulletin. 2021;**11**(2):274-285. DOI: 10.34172/apb.2021.040

[22] Chen Q, Jiang H, Jin P, Zhang J, Wang W, Wang K, et al. Orange fragrance with sustained-release properties prepared by Nanoethosomal encapsulation of natural Orange essential Oil. Journal of Nanoscience and Nanotechnology. 2019;**19**(11):7464-7470. DOI: 10.1166/jnn.2019.16713

[23] Silva P, Bonifácio B, Ramos M, Negri K, Maria Bauab T, Chorilli M. Nanotechnology-based drug delivery systems and herbal medicines: A review. International Journal of Nanomedicine. 2013;**9**:1. DOI: 10.2147/IJN.S52634

[24] Jain S, Tripathi S, Tripathi PK. Invasomes: Potential vesicular systems for transdermal delivery of drug molecules. Journal of Drug Delivery Science Technology. 2021;**61**:102166. DOI: 10.1016/j.jddst.2020.102166

[25] Santos JS, Barradas TN, Tavares GD. Advances in nanotechnology-based hair care products applied to hair shaft and hair scalp disorders. International Journal of Cosmetic Science. 2022;**44**(3):320-332. DOI: 10.1111/ics.12780

[26] Santos JS, Tavares GD, Nogueira TB. Vitamin E and Derivatives in Skin Health Promotion. In: Santos JS, Erkekoglu P, editors. London: Intech; 2021. pp. 1-13. DOI: 10.5772/intechopen.99466

[27] Ammar HO, Tadros MI, Salama NM, Ghoneim AM. Ethosome-derived invasomes as a potential transdermal

delivery system for vardenafil hydrochloride: Development, optimization and application of physiologically based pharmacokinetic modeling in adults and geriatrics. International Journal of Nanomedicine. 2020;**15**:5671-5685. DOI: 10.2147/IJN.S261764

[28] Ahmed OAA, Badr-Eldin SM. Development of an optimized avanafil-loaded invasomal transdermal film: Ex vivo skin permeation and in vivo evaluation. International Journal of Pharmaceutics. 2019;**570**(May):118657. DOI: 10.1016/j.ijpharm.2019.118657

[29] El-Tokhy FS, MMA A-M, El-Ghany EA, Geneidi AS. Design of long acting invasomal nanovesicles for improved transdermal permeation and bioavailability of asenapine maleate for the chronic treatment of schizophrenia. International Journal of Pharmaceutics. 2021;**608**:121080. DOI: 10.1016/j.ijpharm.2021.121080

[30] Ahmed S, Kassem MA, Sayed S. Bilosomes as promising nanovesicular carriers for improved transdermal delivery: Construction, in vitro optimization, ex vivo permeation and in vivo evaluation. International Journal of Nanomedicine. 2020;**15**:9783-9798. DOI: 10.2147/IJN.S278688

[31] Lv X, Cong Z, Liu Z, Ma X, Xu M, Tian Y, et al. Improvement of the solubility, photostability, antioxidant activity and UVB photoprotection of trans-resveratrol by essential oil based microemulsions for topical application. Journal of Drug Delivery Science Technology. 2018;**48**:346-354. DOI: 10.1016/j.jddst.2018.10.017

[32] Souto EB, Cano A, Martins-Gomes C, Coutinho TE, Zielińska A, Silva AM. Microemulsions and Nanoemulsions in skin drug delivery. Bioengineering. 2022;**9**(4):1-22. DOI: 10.3390/bioengineering9040158

[33] Tartaro G, Mateos H, Schirone D, Angelico R, Palazzo G. Microemulsion microstructure(s): A tutorial review. Nanomaterials. 2020;**10**(9):1-40. DOI: 10.3390/nano10091657

[34] Hussein A, Abdel-Mottaleb MMA, El-assal M, Sammour O. Novel biocompatible essential oil-based lipid nanocapsules with antifungal properties. Journal of Drug Delivery Science Technology. 2020;**56**(February):101605. DOI: 10.1016/j.jddst.2020.101605

[35] Shaaban M, Nasr M, Tawfik AA, Fadel M, Sammour O. Bergamot oil as an integral component of nanostructured lipid carriers and a photosensitizer for photodynamic treatment of vitiligo: Characterization and clinical experimentation. Expert Opinion on Drug Delivery. 2021;**18**(1):139-150. DOI: 10.1080/17425247.2021.1844180

[36] Nastiti CMRR, Ponto T, Mohammed Y, Roberts MS, Benson HAE. Novel nanocarriers for targeted topical skin delivery of the antioxidant resveratrol. Pharmaceutics. 2020;**12**(2):108. DOI: 10.3390/pharmaceutics12020108

[37] El-Nabarawi MA, Shamma RN, Farouk F, Nasralla SM. Dapsone-loaded Invasomes as a potential treatment of acne: Preparation, characterization, and In vivo skin deposition assay. AAPS PharmSciTech. 2018;**19**(5):2174-2184. DOI: 10.1208/s12249-018-1025-0

[38] Hussain A, Altamimi MA, Alshehri S, Imam SS, Singh SK. Vesicular elastic liposomes for transdermal delivery of rifampicin: In-vitro, in-vivo and in silico GastroPlus™ prediction studies. European Journal of Pharmaceutical Sciences. 2020;**151**(February):105411. DOI: 10.1016/j.ejps.2020.105411

[39] Tawfik MA, Tadros MI, Mohamed MI, El-Helaly SN. Low-frequency versus high-frequency ultrasound-mediated transdermal delivery of agomelatine-loaded invasomes: Development, optimization and in-vivo pharmacokinetic assessment. International Journal of Nanomedicine. 2020;**15**:8893-8910. DOI: 10.2147/IJN.S283911

[40] Rangsimawong W, Obata Y, Opanasopit P, Ngawhirunpat T, Takayama K. Enhancement of Galantamine HBr skin permeation using Sonophoresis and limonene-containing PEGylated liposomes. AAPS PharmSciTech. 2018;**19**(3):1093-1104. DOI: 10.1208/s12249-017-0921-z

[41] Altamimi MA, Hussain A, Imam SS, Alshehri S, Singh SK, Webster TJ. Transdermal delivery of isoniazid loaded elastic liposomes to control cutaneous and systemic tuberculosis. Journal of Drug Delivery Science Technology. 2020;**59**(May):101848. DOI: 10.1016/j.jddst.2020.101848

[42] Takayama K, Nagai T. Limonene and related compounds as potential skin penetration promoters. Drug Development and Industrial Pharmacy. 1994;**20**(4):677-684

[43] Bolzinger MA, Briançon S, Pelletier J, Chevalier Y. Penetration of drugs through skin, a complex rate-controlling membrane. Current Opinion in Colloid & Interface Science. 2012;**17**(3):156-165. DOI: 10.1016/j.cocis.2012.02.001

[44] Waheed A, Aqil M, Ahad A, Imam SS, Moolakkadath T, Iqbal Z, et al. Improved bioavailability of raloxifene hydrochloride using limonene containing transdermal nano-sized vesicles. Journal of Drug Delivery Science Technology. 2019;**52**:468-476. DOI: 10.1016/j.jddst.2019.05.019

[45] Dsouza L, Chaudhari P, Brahmam B, Lewis SA. Derma roller mediated transdermal delivery of tizanidine invasomes for the management of skeletal muscle spasms. European Journal of Pharmaceutical Sciences. 2021;**165**:105920. DOI: 10.1016/j.ejps.2021.105920

[46] Binsuwaidan R, Sultan AA, Negm WA, Attallah NGM, Alqahtani MJ, Hussein IA, et al. Bilosomes as Nanoplatform for Oral delivery and modulated In vivo antimicrobial activity of lycopene. Pharmaceuticals. 2022;**15**(9):1-23. DOI: 10.3390/ph15091043

[47] Altamimi MA, Hussain A, Alshehri S, Imam SS, Alnemer UA. Development and evaluations of transdermally delivered luteolin loaded cationic nanoemulsion: In vitro and ex vivo evaluations. Pharmaceutics. 2021;**13**(8). DOI: 10.3390/pharmaceutics13081218

[48] Pereira I, Zielińska A, Ferreira NR, Silva AM, Souto EB. Optimization of linalool-loaded solid lipid nanoparticles using experimental factorial design and long-term stability studies with a new centrifugal sedimentation method. International Journal of Pharmaceutics. 2018;**549**(1-2):261-270. DOI: 10.1016/j.ijpharm.2018.07.068

[49] Souto EB, Zielinska A, Souto SB, Durazzo A, Lucarini M, Santini A, et al. (+)-limonene 1,2-epoxide-loaded slns: Evaluation of drug release, antioxidant activity, and cytotoxicity in an HaCaT cell line. International Journal of Molecular Sciences. 2020;**21**(4):1-11. DOI: 10.3390/ijms21041449

[50] DiMasi JA, Hansen RW, Grabowski HG. The price of innovation: New estimates of drug development costs. Journal of Health Economics. 2003;**22**(2):151-185. DOI: 10.1016/S0167-6296(02)00126-1

[51] Kakkar S, Kaur IP. Spanlastics — A novel nanovesicular carrier system for ocular delivery. International Journal of Pharmaceutics. 2011;**413**(1-2):202-210. DOI: 10.1016/j.ijpharm.2011.04.027

[52] Souto EB, Macedo AS, Dias-Ferreira J, Cano A, Zielińska A, Matos CM. Elastic and Ultradeformable liposomes for transdermal delivery of active pharmaceutical ingredients (APIs). International Journal of Molecular Sciences. 2021;**22**(18):9743. DOI: 10.3390/ijms22189743

[53] Huynh NT, Passirani C, Saulnier P, Benoit JP. Lipid nanocapsules: A new platform for nanomedicine. International Journal of Pharmaceutics. 2009;**379**(2):201-209. DOI: 10.1016/j.ijpharm.2009.04.026

[54] Pezeshki A, Ghanbarzadeh B, Mohammadi M, Fathollahi I, Hamishehkar H. Encapsulation of vitamin A palmitate in nanostructured lipid carrier (NLC)-effect of surfactant concentration on the formulation properties. Advanced Pharmaceutical Bulletin. 2014;**4**(Suppl 2):563-568. DOI: 10.5681/apb.2014.083

[55] Kejlová K, Jírová D, Bendová H, Gajdoš P, Kolářová H. Phototoxicity of essential oils intended for cosmetic use. Toxicological Vitrology. 2010;**24**(8):2084-2089. DOI: 10.1016/j.tiv.2010.07.025

[56] Rangsimawong W, Wattanasri P, Tonglairoum P, Akkaramongkolporn P, Rojanarata T, Ngawhirunpat T, et al. Development of microemulsions and Microemulgels for enhancing transdermal delivery of Kaempferia parviflora extract. AAPS PharmSciTech. 2018;**19**(5):2058-2067. DOI: 10.1208/s12249-018-1003-6

[57] Xiao Z, Xu Z, Zhu G. Production and characterization of nanocapsules encapsulated linalool by ionic gelation method using chitosan as wall material. Food Science Technology. 2017;**37**(4):613-619. DOI: 10.1590/1678-457x.27616

[58] Trucillo P, Campardelli R, Reverchon E. Antioxidant loaded emulsions entrapped in liposomes produced using a supercritical assisted technique. Journal of Supercritical Fluids. 2019;**154**:104626. DOI: 10.1016/j.supflu.2019.104626

[59] Galiniak S, Mołoń M, Biesiadecki M, Bożek A, Rachel M. The role of oxidative stress in atopic dermatitis and chronic Urticaria. Antioxidants. 2022;**11**(8):1-13. DOI: 10.3390/antiox11081590

[60] Dobrică EC, Cozma MA, Găman MA, Voiculescu VM, Găman AM. The involvement of oxidative stress in psoriasis: A systematic review. Antioxidants. 2022;**11**(2):1-32. DOI: 10.3390/antiox11020282

[61] Ponto T, Latter G, Luna G, Leite-Silva VR, Wright A, Benson HAE. Novel self-nano-emulsifying drug delivery systems containing astaxanthin for topical skin delivery. Pharmaceutics. 2021;**13**(5). DOI: 10.3390/pharmaceutics13050649

[62] Theochari I, Mitsou E, Nikolic I, Ilic T, Dobricic V, Pletsa V, et al. Colloidal nanodispersions for the topical delivery of ibuprofen: Structure, dynamics and bioperformances. Journal of Molecular Liquids. 2021;**334**:116021. DOI: 10.1016/j.molliq.2021.116021

[63] Amnuaikit T, Limsuwan T, Khongkow P, Boonme P. Vesicular carriers containing phenylethyl resorcinol for topical delivery system; liposomes, transfersomes and invasomes. Asian Journal of Pharmaceutical

Sciences. 2018;**13**(5):472-484. DOI: 10.1016/j.ajps.2018.02.004

[64] Chen J, Jiang Q, Ping CY, Zhang H, Peng P, Yang X. Natural terpenes as penetration enhancers for. Molecules. 2016;**21**:1-22. DOI: 10.3390/molecules21121709

[65] Elmowafy M, Shalaby K, Ali HM, Alruwaili NK, Salama A, Ibrahim MF, et al. Impact of nanostructured lipid carriers on dapsone delivery to the skin: in vitro and in vivo studies. International Journal of Pharmaceutics. 2019;**572**(October):118781. DOI: 10.1016/j.ijpharm.2019.118781

[66] Elmowafy M. Skin penetration/permeation success determinants of nanocarriers: Pursuit of a perfect formulation. Colloids Surfaces B Biointerfaces. 2021;**203**:111748. DOI: 10.1016/j.colsurfb.2021.111748

[67] Winstanley P, Orme M. The effects of food on drug bioavailability. British Journal of Clinical Pharmacology. 1989;**28**(6):621-628. DOI: 10.1111/j.1365-2125.1989.tb03554.x

[68] An Y, Park MJ, Lee J, Ko J, Kim S, Kang DH, et al. Recent advances in the transdermal delivery of protein therapeutics with a combinatorial system of chemical adjuvants and physical penetration enhancements. Advances in Therapy. 2020;**3**(2):1900116. DOI: 10.1002/adtp.201900116